THE ORIGIN OF LIFE ON THE EARTH

THE ORIGIN OF LIFE ON THE EARTH

An Historical Account of Scientific Thought
Part I - From the beginning of time until 1900

MK GALI

Scientum Publishers LLC

First Printing, 2020

To every person
who has influenced my thoughts and life
making me a better person than I was
before

PREFACE

Since Darwin first proposed his theory of natural selection, the question of the origin of man has been endlessly debated. So it is natural that the curious mind would follow all the recent developments on this matter. Today, two groups continue to struggle with each other over this question. One group believes in science and the theory of evolution. Another group consists of creationists and young-earth creationists who believe in an intelligent design. And, at least in the case of man's origin, science offers a clear explanation. But, on the question of the origin of life, however, science cannot yet offer a definite answer.

Since the beginning of time, every man on Earth has pondered over the origin of life. For a long time, man's efforts to answer this basic question were in vain. As his collective knowledge kept growing, he came up with more interesting stories and theories. Initially, these were imaginative, later based on some interesting observations, and then only centered on empirical evidence. Until last few of centuries back man was confident that the question of origin would remain unanswered, but now we feel close to generating life from inanimate objects in a lab. This significant shift is not merely due to the private work of individuals but is closely related to changes in social, religious, scientific, and political conditions throughout the world. Understanding this relationship between the changes in the different fields is the key to any historical work. Otherwise, we would learn the ideas, theories, or evidence put forth by the individuals and the right or wrong parts of them, but we would never know "how" or "why."

Scientists, from ancient times to today, are still trying to find an answer for the problem of origin. Many theorists have come up with different, yet to be proved, suggestions. But, science has seen enormous progress in these efforts. The more we understand the structure and function of living organisms, the closer we get to find an answer for the origin of life. Our knowledge on this today, compared to just a century ago, is immense. But unfortunately, unlike evolution, most children today learn no theories related to the origin of life from their textbooks. One reason for this might be the same complaint that there is no definitive answer available in science. Yet, science is not just a study of what has already been found; it is the study of nature regardless of whether we have all the answers to a problem or not.

The problem with keeping non-scientists ignorant of developments is that they will try to find the answers elsewhere. Man's curious mind seeks to fill the void in the explanation of probable cause. In his efforts to do so, he is prone to make errors no matter how strong his reason of argument is. If the void is big, then the probability of erring will also be high. If science has no answer, then other fields like theology or philosophy, which provide an answer backed with strong arguments, begin to look more appealing.

Even though theories about the origin of life are routinely ignored in formal education, aren't there other sources of information related to such theories available for an interested reader? Yes, of course. There are many books on this subject. Scientists directly involved in the study of the origin of life on Earth often write their own texts, and the popular science writers closely following the advances in the subject have composed more accessible volumes. These books present extensive technical details of the recent breakthroughs achieved, interesting problems faced in the research, and the larger meaning inferred for the humanity. There is also another group of authors who compare creation theories with the progress achieved in science. The focus of the first set of authors is to provide us with an account of the current status of the origin of life, while the focus of the second set of authors is on supporting or refuting the creation theory. But despite the breadth of works on the subject, I have failed to find a comprehensive historical account of the problem of the origin of life. The rich history of more than two millennia has thus far been largely ignored, or covered so briefly that we fail to understand the influence of social, political, philosophical, and scientific changes on the progress of the subject.

The historical account is necessary to comprehend why we stand where we are. Otherwise, we risk repeating the same mistakes our predecessors did, or losing focus on the current and future course of the priorities. We may even find many of the historical aspects amusing if we don't have the appropriate background information. For example, it may seem absurd that Jan Baptist van Helmont said that all vegetables gain mass only from water, but it followed a strong philosophical argument and was backed by the first quantitative experiment ever done in biology. Van Helmont followed the most logical and advanced concepts and procedures of the time, and he reached a conclusion that made sense to many of his peers.

In this book, I attempt to provide a comprehensive historical account of the origin of life on Earth, at least as we humans have understood it. I have tried to provide the reader with a broad description of events and the outcomes, which includes the circumstances that inspired the individuals in their opinions or works. I have tried my best to refrain myself from forcing any personal views, so the reader has more freedom to interpret the ideas as he or she sees fit.

One dilemma I had was whether to include creation theories in the book. The thought of adding another part to the book on religious theories tempted me often. But I found that if I included such theories, the book would become simply another "science versus religion" discussion. I want to keep the reader from any such influences, and more importantly, I want to dedicate this book purely to scientific discussions. Details from any other field are discussed only if the scientific community or the individuals involved are affected by the events in those areas.

I have divided this book into two parts for a particular reason. The first one tracks the developments from the classical period to the end of 19th century, and the second covers the progress from the beginning of 20th century to the present. Even though the second part picks from where the first one finishes, the nature of the theories we observe in these two sections differs greatly from each other.

A Final Note

Before starting with this task, I had a question to answer: Am I qualified to write this book? I am neither a scientist nor a historian. But I consider myself a student of science interested in those branches that are relevant to society. Being an activist for science, I believe there is a need to take science to the common man in a way he can understand it and apply the scientific method in his day-to-day life. The reason for taking this task upon myself is my own failed attempt at finding such a work when I needed it. I don't find it necessary to comprehend all the complex scientific equations in a research paper, but I find it crucial to follow the developments in science and its possible applications to social problems. Many significant developments in science that would make society better are still not reachable to a vast section of the population. It is surprising how unscientific traditions persist in our day-to-day life. One such example is Thomas Redi's extensive work on the vipers in the 17th century in which he refuted the myth that snakes drink milk. But even now, to pick an example, in India it is widely believed that snakes drink milk. What is the use of a scientific discovery if the society does not benefit from it?

Things that a scientist might find very obvious could be difficult to grasp for the common man. So I hope this work will bridge such gap between scientists and ordinary citizens by approaching the topic in simple language that a layperson can easily understand. By the end of this book, if the reader can understand the progress in the origin of life on Earth and its implications, then I will consider myself fortunate.

CONTENTS

INTRODUCTION

Life is abundant on Earth. To date, we have identified more than a million life forms, but this pales in comparison to the vast majority of species still undiscovered. The number of extinct life forms is few thousand times more than the total number of life forms that presently exist. Over time, we have observed that the life can survive even in extreme conditions, like in the hottest volcanic springs, below the coldest ice of Antarctica, in the rocks below the sea floor, in barren deserts, in acidic conditions, and in places lacking oxygen. Even a century ago, such statements would be considered laughable.

Among all the existing life forms man takes pride in thinking he is the most intelligent. Man's ability to generate new knowledge and to pass it on to the next generations has given the human species an immense advantage over other life forms. He grew stronger as time passed and successfully exploited nature to his gain. His hunger for knowledge is infinite. He found answers to many complex riddles posed by nature. Even in his efforts towards finding an answer to the origin of life question, man has made many discoveries and come up with theories that have helped shape the different fields in a variety of ways.

One important theory on the origin of life stands apart until the end of 19th century: spontaneous generation. Until the beginning of 20th century, spontaneous generation was the prevailing theory for the origin of lower life forms. Generation of higher life forms, like humans, were not studied exclusively in science. Those who tried to find an answer for the generation of higher life forms mostly relied on either religious texts or spontaneous generation.

The theory of spontaneous generation began because the existence of microorganisms was unknown. When an animal or plant appeared from no visible seed, then the first natural instinct was to attribute it to spontaneous generation. This simple thought led the first investigators to find a recipe for creating each new plant or animal. They tried to set up a controlled environment, and observe how life generates out of non-life, or from a putrefying body. Life appeared, but investigators could only fathom what they could see.

All the investigators were so busy in identifying such recipes that they could not shift their thoughts from it when a breakthrough came from Anton Van Leuwenhoek in the 17th century. Microscopes opened the doors for the new world of microorganisms. But most people, including Leuwenhoek, failed to consider the role of microorganisms in putrefaction or fermentation. Instead, most scholars were more comfortable with the age-old idea than to a fundamental shift in their worldview.

Moreover, many other ideas, in nature very similar to the theory of spontaneous generation, continued to boost it. Vitalism and Miasmata are examples for such theories. Interestingly, none of these theories had any credible scientific evidence in the modern sense, but they continued to support one another. To understand how this worked, let's consider an experiment devised to test a theory 'A.' Investigators would interpret the outcome of the experiment by taking another theory 'B' for granted. If theory 'B' is wrong, then the interpretation of the experiment is erred. It means theory 'A' still has no credible evidence.

Outside the scientific community, theories supporting the origin of life on Earth by natural means found support, or opposition, from scholars in other schools of studies such as theology and philosophy. Theologians and philosophers associated such theories with God. The general line of thought suggested that if the life generated on Earth through a natural process, then it would remove the need for a creator. Few exceptions took a different line of argument suggesting that it was God's wish for the generation of life to happen naturally.

In 18th century philosophy, the idealistic thought became popular, which suggested a teleological principle for all life and opposed any explanation involving a natural means for the origin of life on Earth. But revolutions across Europe by the end of the 18th century resulted in strong support for the materialistic views. Skeptical materialist philosophers supported the theories attributing the origin to a natural means, such as spontaneous generation.

Another proposal for the origin of life on earth went almost unnoticed until the end of 19th century. This theory, now called panspermia, suggested that life exists throughout the universe. From this, an implication was derived suggesting the seeds of life from space, most likely through a meteorite, must have reached earth. The nature of these seeds of life reaching the earth was not discussed by the proponents of this theory in the 19th century. Also, this might solve the problem of the origin of life on earth, but it will still open another question before us, the origin of life in the universe.

As we stepped into the 20th century, spontaneous generation had been rendered irrelevant in the wake of atmospheric germ theory. And panspermia had never gained the popularity that spontaneous generation did. There was a desperate need for new efforts and a new direction to understand how all life began. One of the most debated scientific theories in the history, Darwin's theory of evolution by natural selection, gave this much-needed direction. Darwinism changed the way people looked at the life in general, and it affected everything that came after. Scientists took the idea derived from Darwinism forward and tried to form a strong hypothesis about the beginning of life. The discovery of double helix structure of DNA by James Watson and Francis Crick took this research to a whole new level, as scientists now tried to identify the process leading to the first building blocks of life. These efforts took two distinct directions. One, to identify the chemical process in early Earth conditions that could have resulted in the first building blocks of life. And two, to find evidence of life in space and confirm the method by which it came to early Earth. Both these efforts depended on Darwinism to explain how that first life led to every other life form on Earth.

When we learn all these efforts, one of the major complaints pointing to the lack of a conclusive answer for the big question takes a back seat, and our appreciation for the feasibility of different scientific proposals will only grow.

CHAPTER 1

FIRST ATTEMPTS

The question of the origin of life is so fundamental that we can never be confident enough to know precisely where man's efforts began. While many interesting theories were put forward in ancient religious or mythological texts from Babylon, Egypt, Mesopotamia, India, and China, they all involve the hand of a supernatural entity. The Greeks put together the first philosophical account of the origin of life, so we will begin our examination there.

Some theories from ancient Greek philosophy included divine presence. For example, in Thales's theory where 'everything' comes out of the water, he considered water to be divine. But we still discuss Thales's theory because it suggests a natural element or matter as the primary substance from which everything else came. The causality rules out the hand of the supernatural element in all following explanations.

Basic Elements

Thales of Miletus takes the credit for being the first in many fields. He was the first known Greek philosopher, the first Western scientist, and the first Greek mathematician. He is often called the "father of science", but perhaps the "father of *Greek* science" would be more accurate. Thales (c. 624–546 BC) was from a Greek city called Miletus. None of his direct writings are available now, and most of what we know about him comes from the writings of others. But his reputation speaks for itself. Based on his prediction of weather, Thales reserved all the olive presses in Miletus far in advance, only to sell or rent them later for a higher price. This is perhaps the first known "future" or "option" contract in the financial world. He also correctly predicted the first solar eclipse in the history of Greece on 28 May 585 BC, famously called "the Eclipse of Thales."

Thales tried to explain natural phenomena based on rational reasoning rather than mythology. Based on his observations, he concluded that water is the primary substance and everything that exists came from it. He believed that the Earth was flat and floated on water. Though we don't know how exactly he came to that conclusion, it appears that he observed how a moist substance would dry and turn into different forms of matter, so he applied that logic to Earth.

If Earth came from water, then everything on Earth came from water, which further implies that life too came from water. This notion suggests that the life originated from lifeless or inorganic matter. But Thales himself never delved into the specifics of how life came into being.

One of Thales's students, Anaximander (c. 611–546 BC), had a different theory. In truth, it is a matter of debate whether he was a student of Thales or was a younger teacher at the Milesian school, a school of thought founded in 6th century Miletus, but Thales's teachings nevertheless had an impact on Anaximander. He applied a similar rational process of reflection in the study of nature, but he came to different conclusions.

Anaximander is the first Greek philosopher to have documented his studies, but unfortunately, only one fragment is available now. So once again we must depend on other sources to learn about him and his studies. Anaximander observed that any given element has two properties of opposing nature. For example, "wet" and "dry" are two such properties that cannot exist together. He further noted that water could never be dry as its primary property is wet. This implied that any matter that comes from water must also be wet. This runs counter to Thales's idea that "everything," including dry matter, came from the water.

Instead of water, Anaximander posited that a primordial matter called "apeiron" or "arche"—terms he coined to describe his theoretical substance—was the source of origin for all life. He defined apeiron as the substance where all the elements along with all the properties are unified. He further explained how the basic elements eventually separated, and how the universe was created. In many respects, this idea is not so dissimilar from the push among today's physicists for a "Theory of Everything" that would unite the fundamental forces of nature.

In support of his theory, Anaximander proclaimed that the Earth floats freely in infinite space, as it would need no support in any direction. His vision is similar to a ball suspended in a vacuum or water with equal pressure on all sides.

This is possibly the first time in history that such an idea was proposed—a truly brilliant deduction for the era.

But Anaximander did not stop there. He proposed that life originated while the Earth was still forming, with fish or fish-like creatures emerging in the sea. This idea appears to be the result of his fossil study. He believed that the first animals were trapped inside the bark of tree and that they broke out of that bark as they grew. Anaximander thought that human infants couldn't have survived independently on primeval Earth, so instead a primitive version of man would have stayed inside the mouth of a large fish. We can visualize this by comparing it with ovoviviparity in certain types of frogs and sharks. Here, eggs will remain inside the body of one parent. Once hatched inside the body, they receive nourishment from the egg yolk. When they are ready to survive in outside world, they move out of their parent's body. In Anaximander's theory, once men and women were able to survive on Earth and feed themselves, they came out of the host animal. This could only occur after a period when Earth was dry enough for humans to adapt themselves to the new conditions.

The idea that fish, or fish-like creatures, emerged from the sea and gave rise to man echoes the very basic idea of evolution that Darwin would deduce centuries later. Unfortunately, there is no way to know how Anaximander derived his evolutionary explanation for life on Earth, as there are no surviving accounts of his studies.

Anaximander did manage to pass down his ideas through others, though. Anaximenes of Miletus (c. 585–528 BC) and Pythagoras (c. 570–495 BC) were among Anaximander's pupils in Milesian school. From his observations and logic, Anaximenes concluded that air was the primary principle from which everything else came. In essence, Anaximenes replaced Thales's primary element of "water" with "air," which then gave rise to all things.

Thales, Anaximander, and Anaximenes form the trio of Milesian School philosophers who made a permanent mark in the history of ancient Greek philosophy. Similar to Thales or Anaximenes, most influential philosopher of the time Heraclitus of Ephesus (c. 535–475 BC). The "weeping philosopher" believed that fire, which all other elements transform into and thus must be the cause of everything there is.

Clearly, many early Greek philosophers believed in one primordial element as the source of all life. It was water for Thales, apeiron for Anaximander, air for Anaximenes, and fire for Heraclitus. Empedocles (c. 490–430 BC), from Acragas city in Sicily, took a step further from his predecessors and argued that four classical elements constitute all other matter: water, earth, air, and fire.

Empedocles's theory started with a primitive mixture of four classic elements—water, air, fire, and earth—but he soon built on them to include two powers called "love" and "strife" that control the mixture of elements. Love attracts the elements together, causing them to mix with one another; strife separates them. Love and strife acting on the elements in different proportions led to the creation of all things.

In Empedocles's scenario, when the elements began to mix, various different organisms began to form. And when these organisms got together, strange hybrids appeared: Men with ox heads, oxen with human heads, and animals with both sexes. But they didn't survive because they couldn't adapt to the given conditions. Only those that could adapt to nature's conditions survived, which are the organisms we see today.

Empedocles's theory was a path breaker, and his ideas regarding animal adaptation would prove prescient. Empedocles had big ideas in other areas of study, too. Besides being a philosopher, Empedocles was also a poet, politician, and to a certain extent had knowledge of medicine. His death in circa 430 BCE is also the subject of legend. As the story goes, he jumped into Mount Etna, an active volcano in Sicily, to prove his immortality. It proved just the opposite.

Empedocles's contemporary, Anaxagoras of Clazomenae (c. 510–428 BC) was a man who considered all the wealth or property he had a hindrance to his quest for knowledge. So he gave up all of it and immersed himself in a hunt for wisdom. He lived in Athens for about thirty years where he produced most of the significant works.

There are only a few fragments from a part of Anaxagoras's book survived today. And most of the remaining work is available from other writers. Anaxagoras famously stated:

> *In everything, there is a portion of everything except Nous, and there are some things in which there is Nous also.*

In other words, everything is in everything. This sounds similar to Anaximander's theory of apeiron, but Anaxagoras has a different take on it. To him, there is one mixture that contains everything, and then there is "Nous,"—or the mind—which is the only thing that is independent of that mixture. Nous, which is infinite and pure, segregates the similar things from the mixture into different groups to form everything there is. But a portion of everything remains in each thing. Otherwise, Anaxagoras justifies, how could hair come from what is not hair, or flesh from what is not flesh? When things mix, a new thing forms. When they separate, it disappears.

Anaxagoras further argued that the primitive mixture included innumerable seeds of all living things. Men and all the animals formed from those seeds. The hypothesis that the life exists throughout the universe is today called panspermia [meaning 'all seed'], and Anaxagoras was the first known person to come up with this idea. But his theory went against all orthodox beliefs of the time, and it was not taken lightly by Greek society. Indeed, this was not the only such theory put forth by Anaxagoras. Anaxagoras also tried to explain natural phenomena like earthquakes and eclipses. He stated that heavenly bodies are nothing but stones, and that the Moon has mountains just like the Earth and gets the light from the sun.

All these revolutionary ideas, as well as his close friendship with the Athens ruler Pericles, put him in a dangerous situation. Pericles (c. 495–429 BC) ruled Athens from 461 BC until his death. He was a prominent statesman, a defender of democracy, and a great orator. He gave importance to philosophy, education, arts, and literature. This brought Pericles and Anaxagoras close to each other, and Anaxagoras exerted significant influence over Pericles's thoughts.

Athens reached its pinnacle during the so-called Age of Pericles, but his successes did not shield Pericles from political enmity. Pericles and his companions became targets of many charges, and Anaxagoras was eventually given a death sentence. Pericles defended him vigorously at trial and saved him from the death penalty. If not for Pericles, Anaxagoras could have been our first Socrates! After the trial, Anaxagoras moved to Lampsakos, escaping the death sentence as well as the dangerous epidemic known as the "Plague of Athens," which claimed Pericles's life in 429 BC. It is noteworthy to mention that Hippocrates, the father of Western Medicine, was on the ground fighting the very same plague at the time!

After Anaxagoras, the origin theories did not see any significant work for a while. But there was a clear development in the field of philosophy as well as science. This development reached a whole new level with Aristotle's brilliant works, which included substantial work related to the origin of plants and animals.

Aristotle

Aristotle (384–322 BC) is, arguably, the most influential man who has ever lived. He is also the greatest polymath world has seen. The first true scientist, he mastered and contributed significantly to a vast range of subjects: biology, philosophy, physics, politics, geology, logic, rhetoric... the list goes on. His massive work, the quality of it, and the reputation his name carried made it impossible to question his observations or theories for more than two millennia. His studies did not just include observation and reasoning. Instead, he made a big leap from his predecessors by collecting evidence, experimenting with specimens, and basing his studies on empiricism.

Plato (left) and Aristotle in Raphael's 1509 fresco, The School of Athens

Aristotle was born in Stagirus, a Greek colony in Macedonia. His father Nichomachus was a court physician to King Amyntas III of Macedonia, but he died when Aristotle was still very young. Aristotle's guardian and uncle Proxenus sent him to Athens in 367 BC when he was around 18 years old. In Athens, Aristotle joined Plato's Academy.

Things changed back home in Macedonia while he was learning and teaching at the Academy for nearly two decades. Soon after the death of King Amyntas, his son Philip ascended to the throne. Philip II became king of Macedonia in 359 BC and started his rule with aggressive attacks on other regions. He wreaked havoc in the Greek world by waging wars, frequently against the Athenians. In 348 BC, King Philip asked Aristotle to join Hermias of Atarneus as part of a political strategy. Discontented with the Academy in the wake of Plato's death, Aristotle happily agreed.

Hermias was the ruler of Atarneus, but he was also a friend and contemporary of Aristotle. They had studied together in Plato's Academy and with Hermias's assistance, Aristotle settled in Assos in Asia Minor and started his first philosophical school. He later married Hermias's niece Pythia and had a daughter with her.

During his stay in Asia Minor, Aristotle worked on botany and zoology. He spent considerable time near the islands of Lesbos for his studies and produced famous works such as *History of Animals* and *Generation of Animals*.

The works of the philosophers preceding Aristotle were the products of observation and reasoning. But Aristotle added an empirical element to his theorization, which involved experimental study. He formalized both logic and empiricism. One example was his famous experiment with the chicken egg. He patiently opened the eggs at varying intervals to observe embryo development and noticed that different organs developed at different stages. Based on this experiment, he developed the theory of "epigenesis." His observations were not entirely correct, but one can blame the limits imposed by the bare-eye examination for it.

Empiricism played a key role later on as science advanced and scientific methodology developed. For popularizing the scientific method, Sir Francis Bacon often gets due credit as the "father of Empiricism," but the scientific world has much to thank Aristotle for as well.

As we study Aristotle's biology works, we find that he did extensive work on the generation of plants and animals. He broadly categorized the generation of life as spontaneous generation and non-spontaneous generation. Non-spontaneous generation includes all reproduction as we observe generally. Life generates life – there is no dispute there. But can life generate from the lifeless matter?

Aristotle thought so. Based on his observations, Aristotle concluded that some animals and plants are generated spontaneously from putrefying earth or the vegetative matter. He also came up with general rules, or recipes, for the spontaneous generation of different organisms.

Aristotle noticed that when the clay vessels were thrown into the sea and mud collected around them, oysters would appear where there were hitherto none. Limnostreae, or lagoon oysters, grow where there is slimy mud. Cockles, clams, razor fishes, and scallops generate in sandy places. We can find ascidian, limpets, and barnacles in the hollows of the rocks. Hermit crabs generate from the soil and slime and move into a shell. As with onions where the younger plant buds off alongside the existing plant, some testacea like mussels generate alongside the bigger ones. And some testacea emit a liquid material of which younger ones grow spontaneously. Such liquid is not semen but, like semen, it comes from the same material as the parent testacean did, which implies that the liquid has generative power. This also explains why the younger ones resemble the original ones. So as a general rule, Aristotle postulated, all testaceans generate spontaneously but differ in the form based on the material differences.

With the insects, Aristotle observed that not all generate spontaneously. The ones that generate spontaneously do so in different conditions. Caterpillars come from the green leaves of cabbage, and from the caterpillars then come butterflies. Cabbageworm originates from the cabbage itself. When the water drain off the slime forms, which decays and turns white first, then black and finally red after which ascarids generate in it. From them, gnats are created. Ticks come from couch grass. From cow dung comes grubs, which then give rise to the May Bugs. From horse manure comes grubs that are tiny but reddish in color. These grubs become motionless after a while, then turn into flies. Some insects come from the timber too, like the horse fly. Aristotle also observed insects coming from animal hair or flesh, and from excrement that has already been excreted or still within an animal's body.

Aristotle cut opened the eels and observed that there were no means to release the eggs or the offspring, which confirmed him that they grow spontaneously. As further proof, he noted that when there is no rain eels don't appear, but when the rain comes eels also appear. This suggested that rain water is essential for the generation of eels. When the heat from the sun produces putrefaction of the matter in water, earthworms generate out of that decayed matter. Eels either come from such putrefied matter or grow in the mud and water.

According to Aristotle, even some fishes, but not all, come into being spontaneously. Aristotle also took pains to correct some writers who mistakenly believe all mullet generate spontaneously. He differentiated female and male mullet belonging to many species while agreeing that at least one species of mullet generates spontaneously. This tells us that Aristotle was not the first to have studied the spontaneous generation in his period. But it was Aristotle's extensive work on this subject that is available today, and it represents a big shift in the theory of spontaneous generation toward the scientific as compared to the previous studies that were more philosophical in nature.

Regarding the naked eye observations, Aristotle probably did the best that any human could do under the circumstances. If we take the case of the generation of flies from the manure, he mentions that the grubs were exceedingly minute when they came into being. Imagine the world without microscopes and without the knowledge of any organisms that are not visible to our eyes. In such a world, is it easier to assume that there are other organisms all around us that are not visible to us, or is it simpler to assume that certain organisms that will simply appear under the specific conditions? The latter thought seems far more rational than the former.

Aristotle's studies in the Asia Minor ended after four long and eventful years when he was called upon by Philip II to tutor his son Alexander III (356–323 BC), or, as he would later be known, Alexander the Great. Aristotle took the offer and returned to Macedonia in 343 BC. Aristotle taught Alexander along with other noble children in Mieza, 'Shrine of the Nymphs,' for three years until 340 BC.

After his studies, Alexander took charge of Macedonia while his father waged wars and was away from home. In 336 BC King Philip was killed by his bodyguard, and Alexander, now 20, began his reign. Within next ten years, Alexander expanded his kingdom as no other conquerer had before and remained undefeated until the end.

While Alexander was busy in his campaigns, Aristotle started his school called Lyceum near Athens. His works at the Lyceum were classified into two categories: one for the public and another only for the school. Alexander was not happy when he heard about Aristotle publishing his works for the public, and questioned Aristotle that if everything he was taught is made public then what else is there that he excels others in? Unfortunately for us, all the works that Aristotle published for the public are believed to have been lost in later years.

Alexander, who had left Macedonia to conquer Asia, would never return home. After the victory over King Porus, his exhausted army revolted forcing him to take a return journey. But on the way back, he contracted fever while resting in Babylon, which soon claimed his life. Once Alexander was gone, Athenian politicians started planning for their liberation and took aim at the Macedonians among them. Aristotle became one of the targets. Aristotle's memorial and hymn for his friend and father-in-law Hermias were used as a weapon to attack him. A religious leader, Eurymedon the Hierophant, charged Aristotle of impiety for writing a hymn for Hermias as if he were God. Sensing the danger, Aristotle moved to his family estate in Chalcis before it could get any worse in Athens. Aristotle recalled the death of Socrates under similar circumstances and wrote that he "would not let Athens to commit the sin twice against philosophy."

In 322 BC, less than a year after moving to the family estate, Aristotle died of natural causes.

After Aristotle

Alexander's death in 323 BC marked the end of the classical period and the beginning of the Hellenistic period. His reign gave an opportunity for the different cultures to mix and made it easier for the people—and knowledge—migrate to various parts of the world. But it was impossible for his large empire to continue united after his death. It soon got divided under his generals, leading to three leagues: Antigonid dynasty based in Macedonia, the Ptolemaic Kingdom based in Egypt, and the Seleucid Empire based in Persia. Romans eventually conquered all three, thus marking the end of ancient Greece.

The ancient Greek empire had fallen, but its influence was still strong. Greeks had such a rich history that it was impossible to end it altogether. Greek culture greatly influenced Romans, so people of Greece continued to lead the same lifestyle even under the Romans. Romans carried forward the works of Greek philosophers. After Aristotle, Stoicism and Epicureanism rose to prominence in the field of philosophy. But Aristotle's influence on Romans was immense and can be seen in their works. Two such examples are Virgil and Vitruvius's works.

Virgil (70–19 BC) was one of the greatest poets of Rome. In one of the three major works published, *Georgics*, he describes a ritual called *bougonia* through which bees can be generated by sacrificing cow or bulls. This ritual was practiced by the bee-breeders in ancient Egypt to generate bees from the putrefying matter. The description starts with the way the room is setup particularly for this ritual, how a calf is beaten to death before they close shut the room, and how bees generate from dead calf's putrefying body:

> *...bees have been born from the putrefying blood of a*

slain bullock...First a small room is chosen, straitened down just to serve for this; they confine it by a narrow tiled roof and cramped walls, and towards the four winds add four windows with slanting lights. Then is sought a calf of two years old with horns already curving from his forehead; his double nostrils and breathing mouth are stopped up, spite of all his struggling, and he is beaten to death and the flesh pounded to pulp through the unbroken skin. Thus they leave him shut close, laying under his sides broken boughs and thyme, and fresh sprays of casia. This is done when west winds first ruffle the waters, ere yet the meadows flush with fresh colours, ere yet the chattering swallow hang her nest from the rafters. Meanwhile the humours heat and ferment in the soft bones, and creatures wonderfully fashioned may be seen, at first limbless, but soon they stir with rustling wings, and more and more drink in the delicate air: until like a shower bursting from summer clouds they swarm forth, or like arrows from the quivering bowstring when light Parthian skirmishers advance to battle.

Similarly, works by the brilliant Roman engineer and architect Vitruvius's (circa 80–15 BC) provide another account of spontaneous generation. His work *De Architectura* is the only surviving book on the architecture from the ancient world. Leonardo da Vinci's famous "Vitruvian man" drawing is based on the details of the human body proportions in Vitruvius's work.

Vitruvius accounted for spontaneous generation in his architectural designs. While deciding the direction that the buildings should be facing, Vitruvius mentions the worms that generate in the moist wind in the south and the west:

> *Bed chambers and libraries should be towards the east, for their purposes require the morning light: in libraries the books are in this aspect preserved from decay; those that are towards the south and west are injured by the worm and by the damp, which the moist winds generate and nourish, and spreading the damp, make the books mouldy.*

This demonstrates just how deep-rooted an idea spontaneous generation had become in the society after Aristotle. Consideration of spontaneous generation in Virgil or Vitruvius's works implies that the spontaneous generation was no longer just an idea, but it had become an accepted theory. The possibility of spontaneous generation was no more a question; society had taken it as a fact. And it remained so for about two millennia after Aristotle.

CHAPTER 2

SCIENTIFIC REVOLUTION

Before the rise of Christianity, there was no consensus about spontaneous generation from a religious perspective. But once Christianity had a strong hold on Western society, the theory became more popular since people took religious verses literally. The Hebrew Bible strongly hints at spontaneous generation, like the swarm of bees in the body of the carcass or the insects in the winds. Thus, the theory of spontaneous generation theory persisted for many centuries.

By the 5th century, the Western Roman Empire had fallen, and with it so had the world's interest in new science. But two centuries later, the birth of Islam led to a renewed interest in empirical thinking. The Islamic world grew at an astonishing rate, and by mid-8th century it entered a golden age. Science flourished, and medicine, mathematics, and physics saw significant advancements.

Institutional structure is Islamic world's great gift to humanity. In this era, universities began issuing degrees, hospitals required medical diplomas for doctors, and mental institutions began to treat patients with psychological problems. All these new institutions stressed keeping written documents of their works and those of past societies so that knowledge could be passed on to subsequent generations.

Muslim scholars greatly admired Aristotle's work and followed the works of ancient Greek with keen interest. But unfortunately, very little work was done on the origin of life during this period.

In the mid-13th century, Mongols sieged and sacked Baghdad in 1258 CE, ending the Islamic golden age. The Mongols destroyed most libraries during this siege, including one of the largest repositories of books in the world called 'House of Wisdom'. While progress declined in the Islamic world after the Mongol attack, it flourished in the European world. Works of the Greek philosophers and Islamic scholars were made available in Western Europe through trade with the East. Unfortunately, the Mongol Empire was not just a threat to the Islamic world. After they had laid waste to Baghdad, Mongols clashed with Eastern Europe and absorbed it completely. Western Europe avoided the same fate, yet it faced a different kind of problem. Famine, disease, and social unrest took hold, strangling the late 12th century Renaissance before it could truly spread. It would be another four centuries before the scientific thought would transform European society completely.

While Europe was in the process of recovering from foreign attacks and natural disasters, the Turkish Persianate Empire, better known as the Ottoman Empire, sieged Constantinople, the capital of the Eastern Roman Empire, in 1453. With the end of the Roman Empire in both East and

West, intellectuals migrated to Northern and Western Europe where things were already improving. Adding the ancient literature back into the European world triggered a revolution in science by the 16th century.

Nicolaus Copernicus's book *De revolutionibus orbium coelestium* (On the Revolutions of the Celestial Spheres) was published in May 1543. A few days later, Copernicus died from a stroke. But the book proposed a revolutionary theory that the world would remember beyond his lifetime: heliocentrism. This theory, suggesting that the Earth revolves around the Sun, was hard for Copernicus's contemporaries to understand and accept. And it faced the wrath of church and was later banned. Rightfully, the publication of this work marks the beginning of the scientific revolution.

The scientific revolution demanded experimental evidence to support any theories. Rather than trusting just logical reasoning, more stress was now given to devising the experiments to support a particular claim. One of the first and finest experimenters in this era, Joan Baptista van Helmont (1580–1644), came up with some recipes for the generation of certain life forms.

Experimenter – Mystic

Van Helmont was a Flemish chemist and physician, is known for coining the term "gas." He studied how gas forms from dried grapes, and how the wines of grapes, apple, and berries produce gas upon heating. From this study, he confirmed for the first time that gas differs from the air.

Van Helmont practiced medicine as a profession, but he

had also studied philosophy with keen interest, which led him to approach the problems of science as well philosophy with the same fervor. His study of philosophy further pushed him towards the subjects of mysticism and alchemy. Van Helmont's works are collected and published by his son Franciscus Mercurius van Helmont in the book *Ortus Medicinae* (The Origin of Medicine) in 1648. And in these works, we can see that Van Helmont's philosophical, religious, and mystic ideas influenced his scientific experiments.

Willow Tree Experiment

Van Helmont studied the Holy Scriptures to determine the truthfulness of four classical elements theory put forth by ancient Greek philosophers. According to this theory, water and air existed even before creation. God created earth out of the water, which became the third element. But the creation of fire is never discussed in the Bible, so Van Helmont denied to accept fire as an element altogether. Thus, he rejected the theory of the four classic elements and instead proposed that there are only two primary elements, air and water. Earth comes from water, so it is by definition a secondary element. Taking this further, if air and water are the only two primary elements, then everything that is visible must have come from water, and everything that is invisible must have come from the air.

Van Helmont also suggested that even gas is primarily made from water itself. He even experimented by keeping coal submerged in water inside a hermetically sealed flask. The coal, which exhales gas when subjected to fire, dissolves completely in the water after a few days. Thus, Van Helmont concluded that even gas is nothing but water, confirming his theory.

Similarly, Van Helmont believed that all the vegetables, being visible entities, must have come from water. To test this, he conducted his famous willow tree experiment over a period of five years.

Van Helmont took about 200 pounds of dry soil in a furnace and rooted a five-pound willow tree trunk in that soil. After that the only thing he added to the vessel was water. After five years, he took the tree out and weighed the dried soil. There was no change in the soil weight, so Van Helmont concluded that the tree that weighed about 169 pounds now did completely so from the water intake.

> *...I have learned by this handicraft-operation, that all Vegetables do immediately, and materially proceed out of the Element of water onely. For I took an Earthen Vessel, in which I put 2oo pounds of Earth that had been dried in a Furnace, which I moystened with Rain water, and I implanted therein the Trunk or Stem of a Willow Tree, weighing five pounds; and at length, five years being finished, the Tree sprung from thence, did weigh 169 pounds, and about three ounces: But I moystened the Earthen Vessel with Rain-water, of distilled water (alwayes when there was need) and it was large, and implanted into the Earth, and least the dust that flew about should be co-mingled with the Earth, I covered the lip or mouth of the Vessel, with an Iron-Plate covered with Tin, and easily passable with many holes. I computed not the weight of the leaves that fell off in the four Autumnes. At length, I again dried the Earth of the Vessel, and there*

> *were found the fame 200 pounds, wanting about two ounces. Therefore 164 pounds of Wood, Barks, and Roots, arose out of water onely.*

One shortcoming of this study is that Van Helmont ignores the two-ounce difference in the weight of soil before and after the experiment. The description of the whole experiment is quite short, and he does not provide the reader with any specifics. We are expected to believe that he took extreme care in confirming the correctness of all the steps and measures. Likewise, Van Helmont's lack of understanding of the process of photosynthesis—a term that would not be coined until 1893—affected his interpretation of the outcome. With the available evidence at hand, Van Helmont actually reached the most plausible conclusion from the experiment. Despite its shortcomings, the willow tree project became the first quantitative experiment designed to follow the scientific method in the modern sense.

Fermentation

Van Helmont discussed the ferments in great detail. For him, fermentation was the cause for all forms of transmutation, and he accused all schools of studies of being ignorant of the process. In the case of beer, grains transform into liquid by the act of ferment. Even blood only functions because of fermentation within the "*cook-rooms*" of the body.

Van Helmont said there were two ferment types. One includes a seminal "*Archeus*" within it, which is a "seminal air" flowing within the "living soul." The generation of any vegetable or animal from its parents would always use the first type of ferment. This is because the living organism inspires the archeus and the seed within its body, so they act together to produce a child with a likeness to its parent organism. The other type of ferment will not have the archeus, which implies that it can set a transmutation in motion, but cannot generate the life alone. The odor of the ferment provides the idea or image of the thing it is supposed to generate. All the *irregular* generations, where one form generates from something else, would involve the second type of ferment.

Van Helmont gives an example of the mice generation using wheat and a dirty shirt. Put a dirty shirt into a vessel containing wheat grains, and mice will appear in about 21 days.

> *...if a foul shirt be pressed together within the mouth of a Vessel, wherein Wheat is, within a few dayes (to wit, 21) a ferment being drawn from the shirt, and changed by the odour of the grain, the Wheat, it self being incrusted in its own skin, transchangeth into Mice*

The dirty shirt contains a ferment with a vital strength within it, and the odor from the wheat grains provides a direction to begin the generation. Van Helmont finds it wonderful that the mice generated are not small but are fully formed with distinct sexes, and they further propagate their kind. Similarly, Van Helmont suggests that even the lice, flies, gnats, and worms generate using this latter type of ferment.

Moving to the putrefying matter, Van Helmont suggested that whenever a matter develops putrefaction, from odor, it leads to the generation of worms. Scorpions from basil is an example of the generation of life from putrefaction. Living basil contains an odor enclosed within it, which is used in the reproduction process. But if the basil undergoes the putrefaction process, it changes the nature of odor, so instead of producing basil, it create scorpions. The recipe follows like this: dig a trench in the ground and put damaged or "dead" basil in it. Place the basil tightly between bricks so that no insect or animal cannot enter through them. Finally, expose this setup to the sun. As Van Helmont writes:

> *For neither is it a fiction; but in very deed, the Herbe being bruised, and depressed between Bricks, and exposed to the Sun, Aquitane after some dayes, hath yielded unto us, Scorpions. But the more curious one will say, That the Scorpion came from without, to the sweet smell and food of the Herbe: but that doubt is prevented. For truly, the two bricks being mutually beaten together, did suitably touch each other; so that they hindered the entrance of the Scorpion, as well by their co-touching plainness, as by their weight : But a trench did contain the Herb in the middle.*

Van Helmont stressed on the criticality of ferment in not just the generation of life, but in every possible transmutation. All visible things are made of water, so transmutation, or change from one form to another, is seen everywhere. And without fermentation, it would not be possible. So, Van Helmont asked all the schools of studies to consider the study of ferment that had been ignored thus far.

Another unique contribution Van Helmont made to the theory of spontaneous generation was the concept of the odor. Generation of life from putrefaction was an age-old belief. The widely accepted miasma theory suggested that when a body undergoes putrefaction process, particles called *miasmata* would emanate from the body and spread into the air. These *miasmata* are identifiable by the foul smell and are the reason for spreading disease.

Weapon-Salve Theory

Francis Bacon and Galileo were Van Helmont's contemporaries. Francis Bacon (1561–1626) laid a firm foundation for the modern scientific method. Galileo Galilei (1564–1642) is known mainly for his support for the heliocentric system and the astronomical studies. When Copernicus published his work advocating the heliocentric system, it did not initially generate much controversy. One reason for this is that it was published shortly before his death, so there was no one to accuse and punish. Another reason was its failure to create a sensation among the general public. But Galileo's exemplary work in providing evidence to support the theory led to charges of heresy, even though he had influential friends in power. The brilliant way with which Galileo defended himself saved him from any severe punishment. He was let go with just house arrest.

Just about two years after Galileo's trial in Italy, Van Helmont was charged with heresy in Brussels. The reason was his love for mysticism. He was an ardent follower of Paracelsus (1493–1541) and referred to his works in most writings. Paracelsus was a chemist and physicist but was also strongly associated with occult philosophy. Paracelsus contributed a great deal to the world of chemistry and medicine, especially toxicology. He even coined the term "zinc." But at the same time, he was popularly called "the devil" for his interest in the occult. Paracelsus claimed that he had learned a method to cure the wounds during one of his journeys. According to this method, when a special ointment is applied to the weapon that caused a wound, then the wound would heal the wound itself. This procedure, called "weapon-salve," started a hot debate in the chemistry world.

Most of the eminent scholars of the day were involved in the weapon-salve debate. Robert Fludd, an English physician who was also a follower of Paracelsus and known for his involvement in many controversies, defended the weapon-salve theory aggressively. Then a war of words started between a German philosopher Rudolph Goclenius and Flemish Jesuit writer Jean Roberti, which later turned into a pamphlet war. Goclenius defended weapon-salve, while Roberti called it an evil magic. Van Helmont, too, published a pamphlet defending weapon-salve. He didn't just support Paracelsus's theory but extended it to state that the cure of the wound depends on a natural magnetic link between the blood left on the weapon and the blood inside the wounded man.

Those who attacked the weapon-salve theory accused its supporters of black magic and witchcraft. The nature of such allegations was severe at the time due to political and social conditions. A power struggle between the major religious groups had recently resulted in the Thirty Years War (1618–1648) that claimed millions of lives in central Europe. In the aftermath of war, disastrous famine and disease claimed even more lives. Caught up in the religious strife, many people attributed these catastrophes to things like black magic and witchcraft.

In 1634, the church took Van Helmont into the custody on the charge of heresy after he published his "monstrous pamphlet." Though he was a Catholic, his writings and the association with Paracelsus's works made him suspicious. He was released after few days due to lack of the evidence and placed under house arrest for at least two years. Still, he had to suffer allegations by the church for the rest of his life.

Vulgar Errors

During Van Helmont's final years, in 1643, an obscure Englishman wrote a book on Christian faith that made him famous all over Europe. *Religio Medici* (Religion of a Physician) was the first book written by Sir Thomas Browne (1605–1682), an English polymath. The subject matter of this book presented some very sensitive and unorthodox views on religion, and considering the religious strife already going on in the region, Browne quickly drew a lot of attention. But Browne was not in any danger, as he didn't question his Christian faith in the book.

Thomas Browne was born in London on 19 October 1605. His father, a silk merchant on, died when Browne was still very young but left his son a considerable inheritance. By 1637, Browne started practicing medicine in Norwich, where he settled. He was modest, humble, sincere, easy to speak to, and famous for his generous nature towards the poor. All these qualities made him not just an excellent physician, but also a man who respected the opinion of others, no matter what they believed.

Three years after the official release of the edited *Religio*

Medici, Thomas Browne published his next similarly controversial masterpiece *Pseudodoxia Epidemica* (or Enquiries into very many received tenets and commonly presumed truths). Commonly referred to as *Vulgar Errors*, the book lays out Browne's views on superstitions and pseudoscientific claims. Browne dedicates different chapters for different subject categories like animals, man, minerals, plants, deductions from the Holy Scriptures, and picture representations. References to spontaneous generation abound in the book. Browne picks popular beliefs related to spontaneous generation, like Helmont's procedure for the generation of mice, and provides a reason to doubt them. But Browne sticks to the subject of his book strictly and deals only with specific cases, not attacking the theory of spontaneous generation itself. In fact, he expresses his belief in the spontaneous generation in many occasions. For example, he accepts spontaneous or equivocal generation as a possibility from putrefying matter belonging to different animals. He also specifically addressed the generation of worms within the body of man:

> *"...It is not indeed impossible that from the sperm of a Cock, Hen, or other Animal, being once in putrescence, either from incubation or otherwise, some generation may ensue; not univocal and of the same species, but some imperfect or monstrous production, even as in the body of man from putrid humours, and peculiar ways of corruption; there have succeeded strange and unseconded shapes of worms; whereof we have beheld some our selves, and read of others in medical observations. And so may strange and venomous Serpents be several ways engendered; but that this generation should be*

regular, and always produce a Basilisk, is beyond our affirmation, and we have good reason to doubt." – Book 3. Chapter 7.

Though Thomas Browne doesn't refute the theory of spontaneous generation, his questioning mind refused to take the millennia old traditional beliefs for granted.

Faith and reason are the opposite sides of a coin. Man subconsciously blocks reason from questioning his faith. Thomas Browne's religious faith caused him not to question certain superstitions, like witchcraft. At the time, witchcraft was accepted by the majority of the population as a fact of life, and many supposed witches were prosecuted for their magical crimes. At one point, in 1664, Thomas Browne testified in court that two accused girls, Amy Duny and Rose Cullender, appeared to be bewitched. Both girls were subsequently given a sentence of death by hanging. The fact that Thomas Browne was by all accounts a kind-hearted man indicates how firmly he believed in witchcraft.

Ex Ovo Omnia

Not all of Thomas Browne's contemporaries accepted such theories at face value. Another prominent contemporary English physician was more assertive than Browne in denying the theory of spontaneous generation. William Harvey, born on 1 April 1578, was older than both Van Helmont and Thomas Browne. But his work on spontaneous generation was published in 1651, five years after Browne's *Pseudodoxia Epidem-*

ica did. Today we remember William Harvey as the first man to explain the circulation of blood in great detail.

Harvey excelled in his studies and quickly became a physician in charge at St Bartholomew's Hospital. By 1618, he was appointed "Physician Extraordinary" to King James I. Ten years later, in 1628, he published his work *De Motu Cordis* (On the Motion of the Heart and Blood) which directly rejected a long-held view that blood came from the liver.

The progenitor of that view, Galen of Pergamon (129 AD), was a Greek surgeon and physician in the Roman Empire who made tremendous advancements in the field of medicine after Hippocrates. His works were followed and taken for granted for more than millennia, until Harvey directly challenged them. Traditional physicians and academics criticized Harvey, but his reputation as a physician was just too massive to take a hit.

Harvey was riding high among his colleagues until the English Civil War broke out in 1642, which would eventually end absolute monarchy and give birth to the parliamentarian monarchy. As "Physician Extraordinary" to King Charles I, Harvey suffered the consequences of his royal position. When parliamentarians vandalized his house in Whitehall, they destroyed his massive work on the generation of insects and descriptions of the dissections. As a result, he slipped into depression and lost interest in repeating the extensive work all over again. Finally, more than a decade after, he managed to complete his work on insects for the second time.

Harvey showed his draft work to his good friend and anatomist George Ent, who encouraged him to publish it. But Harvey was reluctant as he felt the draft work was not complete, which is most likely because he was not satisfied with the level of work he had done. Finally, he gave Ent permission to publish it in 1651 as *Exercitationes de Generatione Animalium* (Essays on the Generation of Animals). In the book, Harvey discusses the spontaneous generation of insects. Browne's book and Harvey's book were very different. While Browne expressed his doubt on the commonly held use cases of the spontaneous generation based on his inquisitive scientific logic, Harvey denied many cases of the spontaneous generation outright based on his direct experimental works and observations.

In the book, Harvey points to his study of embryo development. Just like Aristotle, Harvey opened the chicken eggs at different time intervals to observe the stages of embryo development. But with a magnifying lens at his disposal, he had a significant advantage over Aristotle. Lens observation led him to discover "cicatricula," which is a point in the egg yolk containing the embryonic cells from which the embryo develops. He further gives an account of "embryogenesis" thus rejecting the related views by Aristotle, Galen, and even his former teacher at the University of Padua Hieronymus Fabricius. Harvey's book further claims that maggots and worms come from their respective eggs and do not, as it was widely believed, generate spontaneously. This led him to famously declare "*ex ovo omnia*"—or, "all life from eggs." For this reason, he is often called the first person to reject the theory of spontaneous generation. But, like Thomas Browne, Harvey pointedly refused to rule out spontaneous generation as a theory, instead attacking those specific cases related directly or indirectly to his work.

Thomas Browne was a great writer, theologian, and philosopher who happened to be an excellent physician with a scientific mindset, but Harvey did not involve himself in different fields of studies other than being a great physician whose works were based solely on his studies. Harvey, unlike Browne, was a skeptic of witchcraft. Working close to the king, he was also involved in some of the witchcraft trials and often helped the defendants win acquittal. In 1632, he dissected a toad said to be used by a witch and confirmed that it was just an ordinary toad with no supernatural ability.

Both Thomas Browne and William Harvey received heat from the conservative section of the society. The same year Harvey's book *Exercitationes de Generatione Animalium* was published, Alexander Ross (1590 – 1654 published his book *Arcana Microcosmi, or the Hid Secrets of Man's Body discovered, in Anatomical Duel between Aristotle and Galen; with a Refutation of Thomas Browne's Vulgar Errors, from Bacon's Natural History, and Harvey's book De Generatione*. In this book, Ross criticized both Browne and Harvey's works. Alexander Ross), a Chaplain-in-Ordinary to King Charles I, was a staunch opponent of any revolutionary ideas. It is ironic that he lived during an era of scientific revolution that gave birth to the revolutionary ideas relentlessly. Ross was fated to be proven wrong no matter how passionately he defended the traditional ideas. Ross supported Aristotle and Galen's orthodox works as a faithful follower. And so he criticized most of his contemporaries who came up with a new science that directly conflicted with Aristotelian or medieval science. He said the following about Thomas Browne in his book:

He doubts whether mice can be procreated of putrifaction. So he may doubt whether in cheese and timber worms are generated; Or if Betels and wasps in cowes dung; Or if butterflies, locusts, grasshoppers, shelfish, snails, eels, and such like, be procreated of putrified matter, which is apt to receive the form of that creature to which it is by formative power disposed. To question this, is to question Reason, Sense, and Experience. If he doubts of this, let him go to Aegypt, and there he will finde the fields swarming with mice begot of the mud of Nylus, to the great calamity of the Inhabitants. What will he say of those rats and mice, or little beasts resembling mice, found in the belly of a woman dissected after her death, of which Lemnius is a witness, who thinks this generation might proceed of some sordid excrement or seminal pollution of those animals with which the womans meat or drink had been infested. I have seen one whose belly by drinking of puddle water, was swelled to a vast capacity, being full of small toads, frogs, evets, and such vermin usually bred in putrified water. A toad hath been found in a sound piece of timber.

William Harvey: frontispiece to his book Exercitationes de generatione animalum. Reproduction of an engraving.
Wellcome Collection. Attribution 4.0 International (CC BY 4.0)

William Harvey passed away in 1657, and Thomas Browne moved on from *Pseudodoxia Epidemica* and published two more philosophical works by 1658. In 1665, about a year after he had given an opinion at the witchcraft trial, he got involved in the struggle against the Great Plague of 1665–66 that claimed more than 25% of London's population.

London's ill fate was far from over even after the Great Plague. About six months later, the Great Fire of 1666 left the 75% of the city, including the St Paul's Cathedral, into ruins. Except for the noble and the lucky few who had managed to save their belongings, the city's inhabitants were plunged into misery and poverty. After the Great Fire, Robert Hooke (1635–1703) , who was the Surveyor to the City of London, worked with his close friend Christopher Wren, a great English architect, to rebuild the city.

This was the same Robert Hooke who coined the term "cell" in biology. Another great polymath of the 17th century, his vast interests included mechanics, architecture, biology, astronomy, and art. Born on the Isle of Wight, he moved to London in 1648 after the death of his father. He eventually became the assistant to Robert Boyle, who is famous today for the law of gas pressure that bears his name. Hooke worked as Boyle's assistant for about seven years, until 1662, when he got an opportunity to join the Royal Society as a curator.

The Royal Society, founded in November 1660, held regular discussions in which Robert Hooke participated regularly. Hooke served as the Society's curator until 1677, and later took the role of physical sciences secretary until the end of his career in 1683. Hooke demonstrated numerous experiments before the Royal Society, which included his microscopy observations. In 1665 he published these microscopy observations in the *Micrographia: or some Physiological Descriptions of minute bodies made by magnifying glasses with observations and inquiries thereupon*.

Robert Hooke's microscope was a much better instrument than William Harvey's small magnifying glass. Indeed, Robert Hooke was more interested in the microscope itself than any particular subject. He studied numerous objects under the microscope with a significant precision and observation. Hooke made notes on some cases of spontaneous generation in Micrographia, including this passage where Hooke notices mites generating from eggs and not spontaneously:

> "*...it seems probable, that some kind of wandring Mite may sow, as 'twere, the first seeds, or lay the first eggs, in those places, which Nature has instructed them to know convenient for the hatching and nourishing their young;... And from some such accidents as these, I am very apt to think, the most sorts of Animals, generally accounted spontaneous, have their origination, and all those various sorts of Mites, that are to be met with up and down in divers putrifying substances, may perhaps be all of the same kind, and have sprung from one and the same sort of Mites at the first.*" - Observ. LV. Of Mites.

Micrographia - Hooke, Robert, 1635-1703

Hooke refuted spontaneous generation only where he could directly observe evidence that was contrary to popular beliefs. Like others of his era, he took spontaneous generation for granted in all other cases. In *Micrographia*, we find more references for spontaneous generation than against it. For example, while discussing the moss or mold-like organism on the blighted leaves of rose shrubs, he supports spontaneous generation:

> *"...the parts of the leaves grow scabby, or putrify'd, as it were, so as that the moisture breaks out in little scabs or spots, which, as I said before, look like little knobs of a red gummous substance.*

From this putrify'd scabb breaks out this little Vegetable; which may be somewhat like a Mould or Moss; and may have its equivocal generation much after the same manner as I have supposed Moss or Mould to have, and to be a more simple and uncompounded kind of vegetation, which is set a moving by the putrifactive and fermentative heat, joyn'd with that of the ambient aerial, when (by the putrifaction and decay of some other parts of the vegetable, that for a while staid its progress) it is unfetter'd and left at liberty to move in its former course, but by reason of its regulators, moves and acts after quite another manner then it did when a coagent in the more compounded machine of the more perfect Vegetable." - Observ. XIX. Of a Plant growing in the blighted or yellow specks of Damask-rose-leaves, Bramble-leaves, and some other kind of leaves.

Similarly, in the case of sea life:

"...for the Sea, among terrestrial bodies, is also a prolifick mother, and affords as many Instances of spontaneous generations as either the Air or Earth." - Observ. XXIII. Of the curious texture of Sea-weeds.

Further, Hooke analyzes other life forms such as mold, mushrooms, worms, and moss and favors spontaneous generation in each case. In all such cases, he finds that the life form generated spontaneously is simpler, smaller, and lower forms of life as compared to the organism they come from.

"...putrifactive Vegetables very often producing a Vegetable of a much less compounded nature, and of a much inferiour tribe; and putrefactive animal substances degener-

ating into some kind of animal production of a much inferiour rank, and of a more simple nature.

Thus we find the humours and substances of the body, upon putrifaction, to produce strange kinds of moving Vermine: the putrifaction of the slimes and juices of the Stomack and Guts, produce Worms almost like Earth-worms, the Wheals in childrens hands produce a little Worm, call'd a Wheal-worm: The bloud and milk, and other humours, produce other kinds of Worms..." - Observ. XIX. Of a Plant growing in the blighted or yellow specks of Damask-rose-leaves, Bramble-leaves, and some other kind of leaves

As far as the cause for such spontaneous generation from putrefying matter goes, Hooke proposes two strong possibilities. The first one involved a "seminal" principle. When some seminal principle, by some insect, is cast on a putrefying matter, it unites with the different putrefying substance and produces various insects or animate bodies. Putrefying matter acts as a nursing place for the generation of an animate body, and the exposure to the seminal principle turns it into a perfect womb.

The second possibility Hooke pointed to involved a vegetative principle. Hooke suggested that there is some vegetative principle in all living beings. When the organism dies, it impedes the vegetative principle's function. Later, as the organs of the body decay further, and as "the heat of the Sun winding up, as it were, the spring," the vegetative principle becomes active again. Hooke believed that a complete living organism constituted a myriad of different parts. Life might be destroyed or ceased because only a few parts stop working, but all other parts would still be intact. This is the reason Hooke thinks that the vegetative principle might function again. The nature of this vegetative principle is different because the complete life doesn't exist. Hence very diverse forms of life can generate from a putrefying body. For the same reason, we may also notice metamorphosis between plant and animal life forms in such cases.

Hooke accepted that the hypothesis of a seminal principle is more *mechanical* when compared to a vegetative principle. But, the vegetative principle here explains the cases where no seminal principle is involved in a spontaneous generation event. Hooke also clarifies that he put forth the vegetative principle hypothesis only as a *conjecture*.

One important point to note here is that the idea of vegetative principle or force was not new. The theory of vitalism, which suggests that all living organisms have some form of vegetative force or vital energy, can be found in many ancient sources. Though Hooke does not refer to vitalism, his understanding of the vegetative principle is in line with the theory.

It is also interesting to note that Hooke did not claim to have witnessed spontaneous generation. Instead, he depends on the findings or writings of other "famous" authors of the time:

> *...at least, if we may believe what is deliver'd to us by very famous Authors; though, I confess, I have not yet been able to discover such myself.*

With the quality of the microscopes at his disposal, Hooke was able to identify the structure of these simple animate bodies, but he could not determine the exact mechanics of their generation. But instead of leaving the question unanswered, he accepted the theory that the majority had favored. One reason for this was that Hooke's primary interest was the microscope itself, rather than the subject of study. He was merely studying these animate bodies using the microscope, but other famous experimenters had dedicated their studies to the generation alone. So Hooke found no reason to question them all. This once again shows how deeply rooted the idea of spontaneous generation was at the time. If one had no evidence against any single claim of spontaneous generation, then he would either accept the spontaneous generation explanation or refrain from denying it.

Another factor to consider is Hooke's religious faith. His father John Hooke was an Anglican clergyman. Hooke wanted to join the Church like his brothers did, but he couldn't because of poor health. Hooke attributes the method in which spontaneous generations work to the creator. An "Omnipotent and All-wise Creator" might have designed the structure and ingredients of such plant or animal, the specific way the putrefying body parts are used in its creation, and how the variations in the degree of heat or other parameters would create different animate bodies. The creator decides if the heating of an egg from fire, bodily warmth, or the Sun should result in which type of organism; and also the mechanism in which it happens,for example an egg of a fly produces maggots, and maggots turn into Aurelia, which will finally transmute into a fly. When we consider the handiwork of the creator in this, we can no longer call such simple organisms imperfect. The Creator provides all the parts necessary for such simple organisms so they can survive and reproduce to propagate their species independently.

In Hooke's view, the creator is akin to a clockmaker. A clockmaker might decide to add a set of chimes to a clock. If we take away the parts of the clock, or the clock stops working, the chimes might still work perfectly. Hooke's view on this may have been influenced by what he saw under the microscope. He was impressed to find simple organisms—both plants and animals—with almost the same structure as any other compound organisms. The only difference was in size.

#

Rich in talents, Hooke still ended up a lonely man in his later years. He grew bitter after being involved in some controversies, initially with Christiaan Huygens and later Isaac Newton. Both Hooke and Huygens invented the balance-controlled watch independently, but the fight came to the claim of who was first. The evidence we now have favors Hooke, but unfortunately, the Royal Society did not support him. Hooke would not let things go without putting up a fight, so he continued with his claim and even accused the then Royal Society secretary Henry Oldenburg of siding with Huygens. Yet most members of the Royal Society backed Oldenburg and rebuked Hooke. Hooke was frustrated by this whole experience, and about a decade later he would spark another big controversy, this time with Newton.

Hooke had worked on the subject of gravity in the 1660s, and he had concluded that the gravity and distance were related by the end of 1670s. Throughout this time, he communicated often with Newton on many subjects as a curator of the Royal Society, and they discussed the matter of gravity as well. Hooke always involved himself into a number of things simultaneously, so he may never have gotten the time to prove his supposition, or probably he just couldn't succeed in this problem. But when Newton published his experimental proof for the same concept in his book *Principia*, he gave no credit to Hooke. Newton acknowledged that Hooke mentioned it to him, but he was not the first from whom he had heard it. Newton might be right because in the 1660s the idea of an inverse square relation between gravity and distance was not exactly new. Nevertheless, things got worse between Hooke and Newton as the time passed. Newton continually refused to give any credit to Hooke, and his responses got harsher as the controversy progressed. At the end Robert Hooke lost the battle, yet again.

Could Newton give Hooke any credit in his book, from a moral perspective? Did Hooke go too far in demanding the credit? It seems like a controversy that could have been averted easily. The contribution made by Newton or Hooke did not die out, and the truth prevailed despite the outcome. This incident is a reminder that, like any other ordinary men, scientists too differ in moral codes.

Two centuries after this controversy, Charles Darwin, and Alfred Russel Wallace would give us an excellent example by dealing with a very delicate situation with immense maturity through mutual understanding and trust. In one of his letters to Hooke, in 1676, Newton says, "If I have seen further, it is by standing on the shoulders of giants." Some writers take this as a way of Newton insulting Hooke by implying that Hooke was not one of those "giants." But the correspondence between Newton and Hooke preceding the use of this phrase indicates their humble nature and respect towards each other's works.

The incident with Newton cost Hooke his reputation. He never married, nor did Newton. Hooke died alone in 1703 leaving all his wealth behind, and Newton became the president of the Royal Society the same year. Newton too had a nervous breakdown at one point, made many enemies, and died alone in 1727. Legend has it that he burned the last surviving portrait of Hooke after becoming the president of the Royal Society, but there is no evidence to support it. Nevertheless, it is true and unfortunate that there is indeed no surviving portrait of Hooke. Many biographers and writers described Hooke as a bitter, jealous, doubting man, but it is possible that they either exaggerated the facts surrounding Hooke's controversies or based their opinion on the views of Hooke's contemporaries who did not like him. But the only way we can truly understand a person is by placing ourselves in his shoes and follow his journey of life.

Other experts in the concerned fields superseded most of Hooke's works. As Newton finished what Hooke was still working on, another brilliant scientist took the microscopy to a whole new level and surpassed the quality of *Micrographia*. But even then, most contributions and observations made by Hooke remain substantial.

Putrefaction

While Hooke was still surveying and rebuilding London after the Great Fire, in Italy, Francesco Redi (1626–1697) made a significant scientific contribution towards the idea of spontaneous generation. Redi was born in Arezzo, Italy and settled in Florence as a head physician at the Court of Medici when he was 22. This is the time when he published his most famous works. Through the experiments he conducted, he aimed to show that spontaneous generation by putrefaction does not exist.

Redi's first major work dealt with the myths surrounding vipers, including common beliefs such as snakes drink milk or wine, or that viper venom is poisonous if swallowed. His next major work in 1668, *Esperienze Intorno alla Generazione degl'Insetti (Experiments on the Generation of Insects)* established him as a leading figure in experimental biology. And his 1685 book *Osservazioni intorno agli animali viventi che si trovano negli animali viventi* (Observations on Living Animals, that are in Living Animals) made him the recognized expert on parasites.

Redi's second book on the insects most directly addresses the theory of spontaneous generation, first by summarizing the works of both the Greek philosophers and his contemporaries. Then he examines each claim one by one and systematically destroys them. He openly mocks a claim that fish thrown out by the sea on the shore of Libya produces worms that give rise to flies, the same flies give rise to the grasshoppers, and finally, grasshoppers change into quails. Then he conducts an experiment to confirm it as false.

As a first step in Redi's experiment, he killed three snakes and puts them in an open box. Shortly after, worms appeared in the box, and their count increased as the time passed. But then, all the worms disappeared. Curious to know where they went, he killed the same type of three snakes again and placed them a closed box. Once worms appeared and all the meat was devoured, they tried to escape but couldn't. He noticed that these worms became motionless over time and turned into eggs. (What Redi called eggs were actually pupae.) Some eggs were red, and some black, and green or black flies came out of those eggs respectively. But this didn't explain how the worms came in the box. Perhaps spontaneous generation?

To find an answer, Redi took six boxes and put two snakes in the first, a pigeon in the second, two pounds of veal in the third, horseflesh in the fourth, a capon in the fifth, and a sheep's heart in the sixth. He also put skinned river frogs into a glass dish. He noticed the same behavior in all six boxes, where maggots appeared first that turned into eggs [pupae] and finally into different flies. For the river frogs, as the dish was open, maggots disappeared after some time.

Repeating the similar set of experiment, Redi took a fish in a box with some holes in it. While maggots appeared in this case as well, he noticed some tiny eggs during the first day. When he opened it again the following day, those tiny eggs were hatched, and maggots had appeared. Once maggots appeared, it was the same cycle all over again. Redi had seen that butchers cover their meat to keep it safe from dirt, and it also reminded him of his colleague's works where flies deposited the eggs on the meat. This made him believe that it was the flies that deposit their eggs first, from which the worms appear. These worms turn into eggs [pupae], which then result in the flies. This must mean that all the worms and flies belong to the same species.

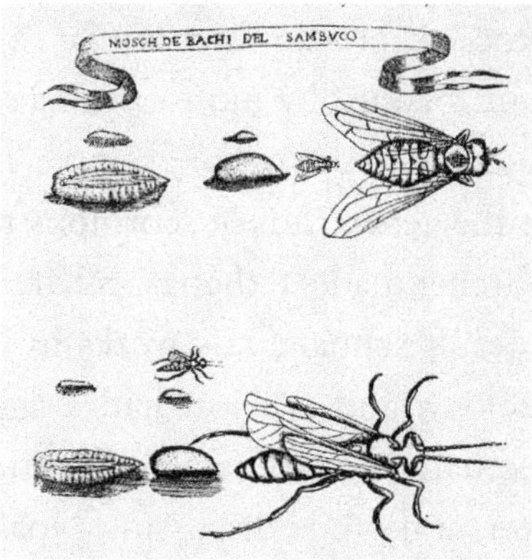

Experiments on the Generation of Insects - Redi, Francesco, 1626-1698

As Redi says, mere belief is vain without the confirmation of experiment. So he decided to conduct his next experiment to confirm his theory. He took four flasks and put snake, fish, eels, and a slice of veal in each of them respectively. He tightly closed all the flasks with a paper cover. He then took a similar set of four other flasks with the same array of meats but left all the flasks open. He noticed later that the open flasks were swarming with worms, but there were no worms in any of the closed flasks. He was not satisfied with just one experiment, so he repeated it with different types of meat in different seasons and found the same result in all of them. He tried burying the meat and still found no worms. He even took dead maggots or dead flies from the meat and conducted the same experiment with the same result. Thus he concluded that the maggots don't generate spontaneously in the putrefying flesh, but they generate from the eggs deposited by the flies on the dead flesh.

Redi then proceeded to conduct similar experiments based on the claims put forward by more famous scholars. In each case, he easily discredited them. For example, he mentions one claim related to the generation of scorpions that his contemporary Father Kircher made. Athanasius Kircher (1602-1680) was a German Jesuit scholar. His works in numerous fields like biology, geology, and religion had a significant impact on many intellectuals of the time. Kircher supported spontaneous generation, and while discussing Noah's ark's dimensions in his 1675 book *Arca Noë*, he left out all the species he thought were generated spontaneously because there was no need to save those species. His take on scorpions was that if dead scorpions were exposed to the sun and sprinkled with sweet basil and water, then that scorpion would generate anew. Redi took this claim and conducted an experiment to verify its truthfulness. After waiting eagerly for the scorpion to reborn, he was disappointed when only maggots appeared instead.

Satisfied with his work on spontaneous generation by the putrefaction of meat, Redi moved on to other sources of this theory. For example, about fruits and vegetable, he came to the same conclusion as for the meat. And while experimenting with cheese, he addressed an idea suggested by Pierre Gassendi (1592–1655). Gassendi was a priest, philosopher, scientist, and astronomer. Gassendi believed that the animals like cattle, while feeding on the leaves, would consume the "seeds" that are deposited the flies. When these animals give milk, it also contains the "seeds," which retain their generative power. This, Gassendi says, explains how worms generate in the milk and the cheese. But, Redi asks, how can the seeds retain their generative power after a process of chewing and digestion? If that were true, Redi adds,

> *one might expect to obtain mullet and eels from cheese made of woman's milk, if the woman had eaten the eggs of the same; or indeed, cockerels and pullets, if she had swallowed hen's eggs.*

Redi further conducted an experiment on cheese to disprove Gassendi's logic. The critical observation and the experimental procedures in Francesco Redi's works are exceptional, and his series of experiments based on causality showcase his impressive deduction skills.

Nonetheless, Francesco Redi believed that the worms from fungus and the worms from fruits or flowers in the trees generate spontaneously. But, in his defense, he did not come to such conclusion hastily. He made observations regarding worms in the trees. For example, regarding the worms from the galls on trees, he supposeed that flies could be hiding their eggs and thus resulting in the galls. Or perhaps somehow the gall appeared first and then flies deposited their eggs. And as always, worms come from these eggs. But then, unconvinced with this line of thought as it seemed to contradict his some other observations, he assumed that the same "natural principal" that leads to the generation of flowers or fruits in the trees must also generate the worms in trees.

Effectively, Francesco Redi refutes spontaneous generation from any putrefaction. Even though he doesn't overrule spontaneous generation ultimately, he stands apart from all his predecessors who denied spontaneous generation to some extent, like William Harvey, because of his well-devised thorough experiments. For the same reason, he shot to fame among his contemporaries.

Religion and Spontaneous Generation

Francesco Redi found many supporters in rejecting the spontaneous generation. Since the beginning of Christianity, the theory of spontaneous generation thrived with few objections. But the scientific revolution began a shift in the opinions. Along with this, the way scholars interpreted the religious texts were also changing.

Because of the scientific revolution, the growth in astronomy and biology was astounding, and many of the great advances were in direct conflict with many widely held views. Any revolutionary discovery in these areas generally puts a question mark on one or the other religious belief. Spontaneous generation also attracted attention as many theologians inferred the support for spontaneous generation as a way of supporting atheism. If the generation of life can happen spontaneously without the predetermined will of the God then why do we need God? This line of thought directed them to refute the entire theory. To take one example, 17th-century naturalist John Ray (1621–1671), who contributed significantly to the fields of fossils, botany, and theology, appreciated Redi's work. He mentions that if one could demonstrate that all forms of life come from parents and not generated spontaneously, then one of the main supporting arguments for the atheism would be taken away. Ray also pointed to the difficulty presented by Redi in attributing the generation of worms in fruits or vegetables to a vegetative soul within the plants. He urged the need to overcome this difficulty to get rid of the theory of spontaneous generation.

Nevertheless, spontaneous generation did not face church's wrath like the heliocentric theory did. There are a few reasons to it. The theory of spontaneous generation predates the beginning of Christianity and had become part of a tradition that most people believed in without ever doubting its validity. Questioning the theory was not taken in the religious angle for the same reason.

Many scientists and the philosophers had taken the related views into a direction that would clash with religion. We have seen the reasons for the same previously. But, none of the experimenters so far had refuted the theory as a whole. Redi, who targeted most claims and came up with evidence against them, still accepted the spontaneous generation in specific cases. So spontaneous generation theory was far from being threatened.

Similarly, even though religious authorities had taken an interest in the subject, they too differed in their opinions. Many favored the notion that spontaneous generation was God's will like everything else is.

Another major reason the debate of spontaneous generation caused no stir in the religious world was that the theory was limited to lower life forms. Spontaneous generation could only be applied to the cases where it was impossible to confirm if the offspring came from its parents. This assured that the theory did not apply to man. As long as the man's creation by God was not questioned, the religious authority tolerated it.

Insect Metamorphosis

While Francesco Redi was preparing to publish his second major work on insects, Jan Swammerdam (1637–1680) was working on a related book called *Historia Insectorum Generalis* (Natural History of Insects), which was published in 1669.

Swammerdam was a Dutch biologist known for his works in the field of anatomy. He studied medicine in Leiden until 1667 but refused to practice. He was far more interested in the study of insects. But his worried father, himself an amateur collector of insects, stopped supporting his son financially, thus forcing him to practice medicine.

Swammerdam made excellent use of his microscope in the study of insects. Swammerdam's well-regarded works on microscopy came just after Robert Hooke's *Micrographia*. Like many others at this time, Swammerdam took interest on the subject of insect metamorphosis. The prevailing thought regarding the metamorphosis was that one particular life form leads to an entirely different life form. For example, Aristotle had discussed how a butterfly comes spontaneously from the caterpillar and how a caterpillar originates from the cabbage leaves. Here, the notion was that the caterpillar and the butterfly are different life forms that generate spontaneously.

Swammerdam, being an anatomist, dissected lot of insects and observed their continuous growth under the microscope at different stages. After microscopic observations of larvae and pupae, including the wings formation in larvae, he concluded that there is no specific point at which a fly comes out of an entirely different life form through some drastic change.

Instead, he suggested that the growth was gradual, just like inside a chick egg.

Among his other works, Swammerdam observed and described the red blood cells for the first time using a microscope. He was also the first to deliver a strong blow to the balloonist theory. Balloonist theory proposed that muscles contract after inflating due to some fluid or airflow. Rene Descartes (1596–1650), who was a French philosopher and mathematician, suggested that the contractions are caused by the "animal spirits" when they enter the muscle. But Swammerdam experimented on frogs and concluded that a simple nerve motion or irritation would cause the contraction, thus refuting Descartes's claim.

Swammerdam's life, though short, was very dramatic. Swammerdam was deeply religious and found it hard to maintain the balance between his spiritual and scientific paths. His love for science was strong, but his spiritual thirst overtook it. Swammerdam's book *The Book of Nature* is a scientific work, but it is full of mystic sentences and poems. His life took a big turn when he came in touch with one of the very influential mystics, Antoinette Bourignon (1616–1680). Antoinette came from a wealthy family who became interested in religion at an early age. She grew very influential over the time with a large following from all sections of the society that constituted scholars, artists, wealthy merchants, and physicians. She prophesied that the Biblical Judgment Day was near and that she—the "New Eve," as she called herself—was chosen by the God to restore Christianity.

Swammerdam had made quite a name for himself among scientists until he came under the influence of Antoinette.

Upon her suggestion, he burned his most famous works, including the one on silkworms. Even though his scientifically inquisitive mind did not vanish fully, his spiritual journey completely overshadowed it. Antoinette died of illness in October 1680, and had Swammerdam lived long enough he might have broken out of her influence and again contributed to science in significant ways. But he died of malaria at the age of 43 in February 1680, eight months before Antoinette's death.

Other Works on the Spontaneous Generation Theory

Before we can proceed further, there are two more important works related to the spontaneous generation theory that needs attention. We will be jumping in time for these studies, but the nature of works are in line with what we have studied so far.

Insect Architecture

James Rennie (1787 – 1867) was a naturalist whose work concentrated mostly on the insects. In 1834 he quit his job as a professor at the King's College, London and for a while relied on his authorship alone. Even before 'Insect Architecture,' his works like 'The Natural History of Insects,' were already familiar. And shortly after publishing the 'Insect Architecture' he moved to Australia in 1840 and started a college high school in Sydney. That is where he settled until his death in 1867.

Like Redi, Rennie too takes on the ancient and contemporary philosophers and naturalists for their views on the spontaneous generation. For example, he mentions one claim supporting the spontaneous generation that said, were a horse's hair put into the ditch water it would first turn into hairworm and then into an eel. This view, which was maintained by even the educated scholars, Rennie says, is 'physically impossible and absurd.' He appreciates the magnificent work done by Redi in the 17th century in discrediting such doctrine while admitting there are still many eminent naturalists who still claim they have witnessed the spontaneous generation under the microscope. Rennie proceeds to refute some of the famous claims for the spontaneous generation in case of insects.

James Rennie refers to a study done by John Murray and team on the honeydew. John Murray and team mentioned that the honeydew was formed spontaneously on the leaves that were washed by sponge and then dried. And the absence of aphides anywhere near the leaf, they claimed, confirms that aphides cannot be the source. But, Rennie says, aphides "are careful to eject the honey dew to a distance from where they may be feeding." He repeated the same experiment with the help of his team by covering leafs by some writing paper. They noticed that the aphides, swarming near the young shoot, ejected the honeydew. As a result, there was no honeydew on the leaf covered with the paper, but the paper itself had honeydew sprinkled all over it within few hours. Rennie also says that they even witnessed aphis ejecting the honeydew with the help of a lens. But, moving further, these aphides are seen appearing simultaneously in great numbers. Where do they come from? This question had led to the widespread notion that winds generate and carry aphides. This, as we have already seen in the Vitruvius's works from the 1st century BC, was a popular notion carried forward in time from mouth to mouth with the little investigation to question its authority. James Rennie dealt with this notion in great detail. Rennie, answering this issue, pointed to the works of two scholars: Reaumur and Latreille.

René Antoine Ferchault de Réaumur (1683-1757) was a great French polymath. He was a naturalist and a physicist. He had done excellent work on aphides and had concluded that one aphis could give birth to about 5904900000 aphides in its lifetime. He further supposes that in a year there could be about twenty generations.

French zoologist Pierre André Latreille (1762-1833), who was trained as a Catholic priest, missed taking the oath of allegiance to the state that was made mandatory for all the priests during the French Revolution where monarchy was already fallen. So he was imprisoned in November 1793 until January 1795. A prison physician took an interest in him when Latreille explained him of the rarity of a beetle he had found in the jail. The physician got in touch with a naturalist named Bory de Saint-Vincent, who luckily already knew of Latreille's works, and managed to get Latreille released. Most of the other prisoners ended up dead soon after, so Latreille would lose his life if he were not released from the prison in time. Latreille took over Lamarck's works at the museum when Lamarck became blind, and even succeeded him as professor of entomology. Latreille had made observations on aphids and had noticed that '*a female during the summer months usually produces about twenty-five a day, each of which will in turn do the same, and so on for several generations in a single season.'*

James Rennie, proceeding with the support of these observations, mentions that his team had seen thousands of aphides on a single leaf of the hop plant, and these swarming aphides in the winter laid 'myriads of myriads' of eggs. These eggs hatch simultaneously in the spring, given they were exposed to same temperature and weather. So those who do not observe the eggs being laid in the autumn will tend to think that the wind carried the insects in spring out of nowhere.

James Rennie further deals with the similar claims of easterly wind carrying the insects. He concludes this problem related to the insects by asking,

It does not seem to have ever occurred to those who thus speak of insectiferous winds, that they get rid of no difficulty by the supposition; for where, we may ask, is the east or any other wind to take up the insects or eggs which it is said to drift along?

Rennie states that the theories, such as winds carrying the insects, arise when the man who is trying to understand how nature operates reaches a point where the established facts are not enough to explain some phenomena; and man's ingenuity makes him fill such gaps with the fragments of imagination. Similarly, as the man couldn't account for the origin of tiny animals and plants, he managed to explain them by coming up with the fancy theories, which collectively led to the spontaneous or equivocal generation theory.

Rennie's work remains one of the landmark works in the history of the spontaneous generation theories related to the insects. There was much similarity between Redi and Rennie's style of arguments, but in the time gap between Redi and

Rennie's timelines, the theory of spontaneous generation had seen a lot of changes.

Fungi and the Spontaneous Generation

Pier Antonio Micheli (1679 – 1737) published the book 'Nova plantarum genera' in 1729 that provided extensive knowledge regarding the fungi. Micheli was from Florence, son of a dyer, who could not afford a formal education. At an early age, he was sent as an apprentice to a bookseller where he learned Latin and studied botany with interest. His excellent hold on the botany subject brought him the position of a junior botanist to Grand Duke Cosimo III de' Medici, Duke of Tuscany. From there he slowly gained name and took multiple responsibilities including a professorship at the University of Pisa.

When Micheli published his 'Nova plantarum genera,' obtaining monetary help from many people, many botanists did not trust the work. It contained a detailed description of about 1400 new plants including the reproduction method in fungi. When he observed that the fungi spores placed on a freshly cut melon would produce a similar type of fungi he thought of spores as any other seeds of plants. Because the seeds obtained from a plant produce the same plant when sown, and spores were no different here. Thus he claimed that the fungi don't generate spontaneously, as the scholars believed, but instead they follow the same mechanism that any other plant does.

About eight years after this work was published, Micheli contracted an infection during one of the specimen-collecting trips that led to pleurisy, and this soon claimed his life on the 1 January 1737.

CHAPTER 3

A WHOLE NEW WORLD

Disproving spontaneous generation in the case of insect metamorphosis is one of the important contributions made by Jan Swammerdam. But three of Swammerdam's friends from his days at school in Leiden—Reinier de Graaf, Nicolas Steno, and Frederik Ruysch—would also leave their mark on the field of anatomy. Their shared teacher Johannes van Horne (1621–1670) was a learned man known for fine anatomical preparations. These five brilliant scientists may have made a more lasting impact on their field, though, had they not gotten involved in petty squabbles over credit for their work.

Swammerdam's friend Reinier de Graaf (1641–1673) settled in Delft as a physician after completing medicine in 1667, and prepared to publish his work on the nature of testis. Having heard about the publication, Van Horne, who was working on the same subject separately with the help of Swammerdam, published his own work on the matter in a hurry. It is not clear if he did it intentionally or not, but Van

Horne left Swammerdam's name out of it, which angered Swammerdam. Then, in 1670, just nearly two years after the incident, Van Horne died, leaving no room for further disputes.

In 1672 De Graaf was ready to publish his next work on the differences between testicles and ovaries. He acknowledged the contribution by Van Horne and Steno but left out Swammerdam's name altogether. Swammerdam claimed that the method of injecting the vessels with the colored fluids for better identification was his own, and he accused De Graaf of stealing it. This led to an aggressive open dispute between De Graf and Swammerdam. Both reached out to the Royal Society to sort out the priority issue. But soon after, in August 1673, De Graaf died at just 32 years of age. De Graaf's cause of death remains mostly speculative. The conjecture is that he committed the suicide under depression following the early death of his son and the stress added by the controversy with Swammerdam. But there is no basis to support these claims.

Few months before his death in 1673, Reinier de Graaf sent a letter of recommendation to the Royal Society asking them to consider the works of one of his friends - Antonie van Leeuwenhoek:

> *...I am writing to tell you that a certain most ingenious person here, named Leewenhoek, has devised microscopes which far surpass those which we have hitherto seen, manufactured by Eustacio Divini and others. The enclosed letter from him, wherein he describes certain things which he has observed more accurately than previous authors, will afford you a sample of his work: and if it pleases you, and you would*

test the skill of this most diligent man and give him encouragement, then pray send him a letter containing your suggestions, and proposing to him more difficult problems of the same kind.

De Graaf didn't survive long enough to know the impact made by his recommendation in the history of science, but humanity is grateful to him for it. Antonie van Leeuwenhoek, though not a scientist, produced such brilliant microscopic observations that it took another two centuries for anyone else to produce anything close to his works.

Little Animals

Antonie van Leeuwenhoek (1632–1723) was five when he lost his father, a basket maker in Delft. He became an apprentice in linen clothes shop in Amsterdam by the time he was 16. Coming from a family with no background in science, he did not involve himself in scientific study. Instead, he opened a linen clothes shop after returning to Delft. He succeeded in the business, and over time, as his reputation grew, he took more responsibilities from the municipal of Delft.

As part of his business, Leeuwenhoek worked on improving the lenses to check the quality of the linen thread as he was not satisfied with the quality of lenses available. This eventually led him to come up with the finest microscope of the 17th century, which remained so for at least another two centuries.

Some unknown reason drew Leeuwenhoek into the field of microbiology. The excited Leeuwenhoek began showing his works to his good friend De Graaf, who immediately recognized their superior quality. De Graaf sent Leeuwenhoek's work and a letter of recommendation to the Royal Society. Henry Oldenburg, the secretary at of the society at the time, recognized that the quality of Leeuwenhoek's works surpassed any other work they had seen before.

Leeuwenhoek did not know English, so he wrote everything in Dutch. His writings also had a peculiar style since he lacked a scientific background. He saw the hand of God in all his works and the more he explored the micro world, the more he wondered about the God's genius mind. Oldenburg himself learned Dutch so he could translate the works of Leeuwenhoek into Latin or English.

For the first time, in 1676, Leeuwenhoek observed single cell organisms, or "animalcules" as he called them, under the microscope. He sent the observations to the Royal Society, which found them hard to believe. A breakthrough in the history of science that would open a whole new branch of study, coming from a man who had no scientific background at all?

The Royal Society asked him to send the improved microscope he was using so they could verify it, which Leeuwenhoek refused. Leeuwenhoek would unfortunately maintain this position not to share the details of his microscopes until the end. As the further proceedings would show he did not want money, but perhaps he was afraid that he would lose the only advantage he had over any other scientist. Since the beginning, he had a notion that all the scientists are learned people who knew much more than he did. His letters reflected his humble nature towards the Royal Society members. So it would be no surprise if he thought that his ideas would not be taken seriously if he made his microscopes public. He did not even keep assistants, as he could trust no one in this matter.

After many independent attempts from the Royal Society members at the verifications had failed, it was Robert Hooke who finally verified it successfully. Some English scholars in Delft, which included some church ministers, visited Leeuwenhoek and confirmed his observations. Robert Gordon, one of those visitors, wrote back to Oldenburg, "*I saw as many as 20,000 of those little animals in a quantity of pepper-water not exceeding the size of a grain of millet.*"

Leeuwenhoek, thus, had discovered bacteria. His work surpassed anything that Robert Hooke had done and published in his *Micrographia*. Referring to the dependency on Leeuwenhoek's works at one point, Robert Hooke apparently said that "*the field had come to rest entirely on one man's shoulders.*" Scholars and dignitaries came from all over to visit Leeuwenhoek and take a peek at his little animals.

Leeuwenhoek refuted spontaneous generation for two reasons. One, he saw God's handiwork even in the new and amazing micro-world he had just found. He couldn't approve of a method of generation by any means other than what was initiated by the God in the beginning. If not from God, then all life must have come from some other parents. The second reason was that his microscopy observations provided evidence against the dominant spontaneous generation claims. For example, he noticed that weevils come out of the tiny eggs rather than emerging from the maize as the people had thought. Similarly, he studied the life cycle of ants, eels, fleas, mussels and many insects to confirm that they all come from their parents.

Epigenesis vs. Preformationism

Leeuwenhoek also discovered spermatozoa, a sperm cell capable of motion and important in the reproductive process. In his observations, he mentions that spermatozoa had many vessels that were nothing but real nerves and veins in a grown up individual body. Here, he was pointing to the preformationism theory.

Aristotle originated the concept of "epigenesis" when he noted that the different parts develop gradually at different intervals inside a chicken egg. By contrast, preformationism says that individual organs do not grow gradually as epigenesis suggests, but instead all the organs are present from the beginning in a very minute form, and they grow slowly over time. Preformationism is easy to understand when we compare it with how a newborn baby grows into an adult. All the parts are already present when the baby is born, and according to preformationism, all the parts will be present even in the eggs or seeds. That is what Leeuwenhoek too supported with his observations of spermatozoa.

One line of thought at the time suggested that if the preformationism theory were true, it would mean that everything comes to life from the seeds or eggs or sperms where the plant or animal already exists as a whole. It means there is no recipe to generate any living organism. This conflicting view would place preformationism at odds with spontaneous generation.

Eel Man and the Vegetative Force

Leuwenhoek opened the doors for a new field of study that

left the whole world dumbfounded. This breakthrough shifted the way people looked at the natural world, and it took a while for people to adjust to the new reality. References to the spontaneous generation of "visible" life forms like insects, worms, and flies continued, but the theory's popularity reduced drastically. Instead, the whole scientific community now shifted its focus to these microscopic life forms.

The progress achieved in microscopy by Leeuwenhoek stalled suddenly after his death. There were no better quality microscopes available, and Leeuwenhoek did not disclose his microscope development secret until the end. As everybody else was using inferior quality microscopes, they could not even question most of his findings or observations. So the work on spontaneous generation theory also slowed down. Finally, it was the works of an 'eel man," which escalated the debate to next level in the 18th century.

John Turberville Needham (1713–1781) was a religious man who pursued biology mainly because of the implications it had on the religious world. So even though Needham received a primarily religious education, he later made quite a reputable name for himself in the scientific community. He was the first clergyman to become a fellow of the Royal Society of London. In one of his papers on geology, he added a section on microscopic observations. He claimed that he had witnessed the generation of eel from rye flour and boiled sheep meat juice, which got him the name "eel man."

In 1748, Comte de Buffon (1707–1788), a rich and famous French cosmologist and naturalist of the time, invited Needham to help with microscopic observations related to his work on the "organic molecules." Buffon had great influence over the following generations with his 36-volume encyclopedic work *Histoire Naturelle*. He himself was greatly influenced by Newton's mechanics and Locke's philosophy, and he sought to keep science free from any religious influence.

Buffon showed Needham "organic molecules" in the reproductive organs of animals. Buffon formulated a theory that the parents contribute these organic molecules when they come together. And similar molecules will amalgamate to form a particular organ of the body. Multiple such groups forming the different organs lead to the whole organism. If there are excess organic molecules, then they will serve as nutrients for the already formed organs. Buffon did not bring God into this theory, but Needham did.

Needham suggested that there is some vegetative force in all such molecules. As God himself willed it, the special vegetative force leads the molecules to come together and fit perfectly to form the organism. He furthered his idea and mentioned that even the Leeuwenhoek's animalcules are formed in the same way and set out to prove it. So he boiled some mutton gravy, placed it in a flask, and sealed it. Later he examined the gravy under the microscope only to confirm his theory as the little animalcules had formed within it. So he concluded that the molecules in the gravy, when heated, organized themselves with the help of vegetative force to form the animalcules. This experiment supported both spontaneous generation as well as epigenesis. Epigenesis versus preformationism was also a matter of hot debate in the 18th century, and Needham was a staunch supporter of epigenesis. He later conducted similar experiments using a different type of matter and came up with the same result.

Spallanzani

No one challenged Needham's claim until an Italian biologist decided to do it 18 years later. Lazzaro Spallanzani (1729–1799) was the son of a reputed lawyer and dreamed of following his father's footsteps by becoming a lawyer. But he took an interest in science under the influence of his relative, Laura Bassi.

In 1608, Juliana Morell, a Spanish nun, became the first woman to receive a doctoral degree. More than a century later in 1732, Laura Bassi became the first female ever to hold a chair at a university. She taught physics at the University of Bologna where Spallanzani went to study law. Under Bassi's influence Spallanzani studied subjects like the mathematics, and after acquiring a taste for science, he never looked back. But unlike Laura Bassi, he moved away from mathematics and found an interest in other subjects like biology. After completing his studies, he worked as a professor and achieved great success. Though ordained as a priest, natural sciences seized his focus. After teaching at various institutions, he joined the University of Pavia in 1768 and worked there until his death. His reputation grew while working at the University of Pavia, and he earned respect from the nobles and students alike.

Spallanzani repeated Needham's experiment in 1768 to verify the truthfulness. He indicated that the first shortcoming in Needham's method was that he had boiled the gravy and let it cool before moving it to a closed flask. By doing so, he argued, Needham had allowed plenty of time for the animalcules to enter the gravy. So he adjusted that experiment by first heating the gravy directly in a flask for an hour while making sure that the air did not enter during or after the process of boiling. Spallanzani also boiled the gravy for a longer time than Needham, over an hour. When Spallanzani tried using other combinations, like letting the boiled gravy open to air or boiling for a shorter period, he found that animalcules appeared. So he concluded that Needham's experiments had important shortcomings, and Needham's claims about spontaneous generation could not be verified.

Unfortunately, this did not stop Needham from arguing that his theory was still right. Needham replied that by boiling the gravy for too long Spallanzani had killed the vegetative force in the molecules. He insisted that Spallanzani's experiments only prove that vegetative force will work only under given conditions. And the other supporters of the spontaneous generation took Spallanzani's experiments in a new direction by claiming that his investigation indicates that air is an essential ingredient for spontaneous generation to occur.

Arguments For or Against the Existence of Spontaneous Generation in Other Fields

The idea of spontaneous generation was a complex one. The experimenters were trying to fight an idea that had originated millennia before through naked eye observations. That idea had evolved over a long period and became part of the tradition. Since there was no well-defined scientific method, it became widely accepted. When scientific study first began, there was no difference between science and philosophy. After the scientific revolution and the development of scientific method, people would be cautious to accept theories without verification. But the trouble with spontaneous generation was that it already had a strong hold within the scientific community. Scholars, still adjusting to the rapid changes in the scientific field, could question use cases of the theory, but not the theory itself. How can you disprove a theory that has no proof but a multitude of cases? Going through every one of such claims and disprove them would be nearly impossible. Yet one cannot simply assume that all the claims in support of the theory are incorrect.

In the 17th century, science was still not mature enough to deal with theories where a far-fetched implication is drawn based on the experimental observations. Needham had introduced the idea of a vegetative force with no actual scientific evidence to support it. His experiment, if conducted with no errors, merely proved that life appeared in the gravy that was boiled within a flask at such and such temperature. To attribute this behavior to a vegetative force has no basis. But the scientific community failed to reject the implication because even the idea of vitalism or an inherent vital force was deep-rooted at the time. Instead, Spallanzani tried to disprove it with the help of another experiment but failed to do so conclusively. So it was impossible to refute Needham's claim unless a strong scientific method or process, which was nonexistent at the time, was followed.

But in the other fields of study, scholars were trying to opine on the matter based on their subject specialties. For example, philosophers rejected a theory if it did not fit their philosophical argument. Theologians accepted a theory only if it fit their understanding of religion. For these scholars, the problem of spontaneous generation theory was more a question of need, rather than of evidence. We will look into some of the important ones here.

Philosophical Arguments or Theories

Voltaire

Spallanzani failed to convince the world that spontaneous generation does not occur. But he had many supporters who approved of his experiment and agreed that it was more accurate compared to Needham's. One of the important supportive figures was Voltaire.

François-Marie Arouet (1694 – 1778), famously known by his pen name Voltaire, was a French philosopher and writer. His writings were versatile and included thousands of pamphlets written using a different pseudonym or anonymously.

Voltaire's dispute with a nobleman resulted in a three-year exile period, which he decided to spend in England. While in England, he compared the political, social and religious differences between the French and the English. Constantly in trouble with the authorities over his critiques of the system, both political and religious, he had to suffer many persecutions and recriminations. But Voltaire did not back from his efforts to reform the system and fight against the inequality, intolerance and absolute monarchy.

Besides his struggles with the French political and religious system, Voltaire also targeted many established personnel in different fields through his satirical style of writing. Voltaire could destroy the reputation of a person systematically with his aggressive mocking style and satirical arguments. During the last decade of his life, Voltaire took aim at the men of science. Needham and Buffon were the targets of his mockery as he considered them to be supporting atheism with their works. He writes about Needham:

> *"A natural philosopher, of some reputation, had no doubt that this Needham was a profound atheist. He concluded that, since eels could be made of rye meal, men might be made of wheat flour; that nature and chemistry produce all; and that it was demonstrated that we may very well dispense with an all-forming God.*
>
> ...
>
> *It is really strange that men, while denying a creator, should have attributed to themselves the power of creating eels.*
>
> ...
>
> *This transmutation of meal and gravy into eels was demonstrated to be as false and ridiculous as it really is, by M. Spallanzani, a rather better observer than Needham. But the extravagance of so palpable an illusion was evident without his observations."*

It is ironic that the man of religion, Needham, who took to science only because he noticed its importance and implications for theology, was considered an atheist by Voltaire because the latter deemed the spontaneous generation to be a threat to the religion.

Immanuel Kant

Immanuel Kant is one of the most influential philosophers of all time, especially in the branches of epistemology, metaphysics, and ethics. And he agreed with Voltaire in rejecting spontaneous generation, but used a slightly different line of argument.

Immanuel Kant (1724–1804) was a German philosopher born in Königsberg, Prussia. His childhood education stressed religion more than science, but his interest in science and philosophy got stronger as he grew up. Along with his philosophical studies, he also worked in the field of astronomy. Immanuel Kant was a critic as well as an admirer of Swedish scientist Emanuel Swedenborg, and he studied most of Swedenborg's works. By 1755, Kant himself had developed the nebular hypothesis, proposed initially in part by Swedenborg, to a great extent. Less than 60 years ago Newton had attributed the order and formation of the solar system to the God. Kant rejected that and explained that cloudy and nebulous material formed the solar system. After this, he immersed himself more and more in philosophical works and earned a name for himself as an influential philosopher.

Kant rejected classical idealism theory and refused to give complete credit to the mind. But he felt the need to identify the relation between lived experience and the ideas independent of experience. He set out to answer these and went into isolation for more than a decade. As a result, he published his long and key work *Critique of Pure Reason* in 1781. But that did not make him famous overnight. He was criticized by most of his contemporaries. It took almost another decade for the world to recognize the importance of his works.

Kant rejected the spontaneous generation of organisms from inorganic matter. He preferred to apply the teleological principle for the living things as opposed to the mechanical principle of the inorganic matter. The teleological principle attributes a purpose or goal for the living organisms. Kant identified three ways in which the generation of living things is attributed in general. The first, "*generatio univocal homonyma*" is when an organism resembles its parents, the most common type of the generation we observe. The second, "*generatio univocal heteronyma*" is when one type of the organism produces another organism that belongs to an entirely different species. While Kant chooses not to refute this idea, he noted that no such examples have been brought to our attention so far. But he ridiculed the third idea of "*generatio aequivoca*," which suggests that an organism generates from completely inorganic matter. He rejected this third type, mostly because attributing a generating purpose to some inorganic matter seemed absurd.

After Kant

Unfortunately, Kant's disciples slowly diverted and used his philosophy to support classical idealism. Kant protested it in vain, for he had moved away from the classical idealism to form transcendental idealism. Kant's philosophy saw a constant decline until the 1860s, when neo-Kantianism would revive it.

After Kant, another German philosopher named Georg

Wilhelm Friedrich Hegel (1770–1831) gained the popularity with dialectical idealism. Though the idea of dialectical method is timeworn going back to the ancient history, Hegel was the first to revive and develop it in the modern philosophical period to form 'dialectical idealism.' In the dialectical method, change results from the inner strife or contradictions. And many philosophers would apply dialectical method to their own philosophical ideas like Hegel did it with the idealism. While Hegel criticized Kant, he still subsumed some of Kant's ideas. But on the spontaneous generation, Hegel accepted it without question because it fit his core philosophical idea.

Goethe

As he agreed with some of Kant's ideas, Hegel also accepted some ideas by another German polymath named Johann Wolfgang von Goethe (1749–1832). Goethe is one of the greatest literary figures in German history. He became famous for his writings in his twenties, but he had also done a lot of work in other areas, especially in science.

Five years after Kant's *Critique of Pure Reason*, Goethe, who had acquired an interest in botany and natural history, conducted experiments related to spontaneous generation. He had put parts of plants in a sealed container with water, which he thought would produce animalcules spontaneously. But his methods were not as advanced as the ones used in Needham or Spallanzani's experiments.

Goethe's works in the areas other than literature and philosophy are generally undermined. This is mainly because his theories could not stand the test of time. But Goethe still influenced many scientists after him. As Ernst Haeckel (16 February 1834 – 9 August 1919), a German naturalist and biologist greatly influenced by Goethe, said:

> *I am sure most of my readers honor Goethe only as a poet and a man; only a few have any conception of the high value of his scientific works, and of the gigantic stride with which he advanced before his own age—advanced so much that most naturalists of that time were unable to follow him. In several passages of his scientific writings he bitterly complains of the narrow-mindedness of professed naturalists, who do not know how to value his works (who cannot see the wood for the trees), and who cannot rouse themselves to discover the general laws of nature among the mass of details. He is only too just when he utters the reproach—"The philosophers will very soon discover that observers rarely rise to a stand-point from which they can survey so many important objects." It is true, at the same time that their want of appreciation was caused by the false road into which Goethe was led in his theory of colors.*

Naturphilosophie

German Romanticism began in the late 18th century, and with it, the *Naturphilosophie* (Philosophy of Nature) was also developed. Fichte, Hegel, and Schelling's philosophical works deal with *Naturphilosophie*. Johann Gottlieb Fichte (1762–1814) admired Kant's philosophy built upon it to trigger the German idealism movement. Kant had initially praised

Fichte's first works but later rejected his other ideas that came in conflict with his transcendental idealism.

Fichte was a mentor of Friedrich Wilhelm Joseph Schelling (1775–1854) for a short time. And Schelling was a roommate and friend of Hegel in the early days. But the friendly relationship between Fichte, Schelling, and Hegel did not survive, as all three disapproved of each other's works over the time. Schelling's works are mostly unknown today partly because of the domination of Hegel's work but also due to the nature of his own work, as it did not provide a complete picture of the philosophical system. Still, he remained a very influential philosopher in the late 18th and early 19th centuries.

Lorenz Okenfuss (1779 –1851), who later shortened his name from Okenfuss to Oken, was a German naturalist and biologist. He extended the works of Schelling on the philosophy of nature and did extensive empirical studies by applying them to the field of biology. He soon became a leading proponent of a kind of *Naturphilosophie* that faithfully followed the views of Schelling. Goethe, who knew Schelling personally, invited Oken to join the University of Jena as a professor in 1807.

In 1805, Oken published a book titled *On Generation*, in which he put forth an interesting theory on the generation of the first living things. He proposed that every organism is nothing but an agglomeration of single cells or mucus mass called the 'mucus vesicles.' These mucus vesicles are also referred as "protoplasma" or "Ur-Schleim" or "infusoria." And Oken further suggested that all the organisms have evolved from these primary infusoria.

Oken published his major work *Elements of Physiophiloso-*

phy in 1810. In it, he attempts to provide a complete structure of nature by applying the ideas from the philosophy of nature. While dealing with the exact mechanism of the generation, he points that every organism consists of infusoria, which are also called primary vesicles. But this begs the question: how did these primary infusoria, which have a life even when they are single cells, form in the first place? They must have originated from the inorganic or else, Oken asks, "*For whence can the organic matter have otherwise proceeded?*"

The concept of polarity is central to Naturphilosophie. Polarity indicates the truly opposed principles united in a body, and nature is full of such principles working together. Oken refers to this Galvanic polarity as the vital force without which there can be no life, and it consists of three vital processes: earth-process (formative), water-process (chemicalizing), and the air-process (electrifying or oxidizing). We can also identify them as the matter, development, and spirit respectively. The primary mucus is generated in the sea and is nothing but oxidized, hydrated carbon. Here carbon is the formative matter that is chemicalized by the water, and then oxidized when it comes in contact with the air. When all three processes combine their energy, mucus is formed. As this creation requires the earth, water, and air to act together, this must have happened in the shallows of the sea when the first mountain peak emerged out of the ocean. For Oken that first mountain peak was Himalaya as he supposed it to be the highest mountain.

Oken called the generation of primary vesicles or organic bodies as the "*Generatio originaria*," or simply "the creation." Every other kind of generation is "*Generatio secundaria*" or

"*Generatio aequivoca*." We must note that the way Oken uses the term "*Generatio aequivoca*" is very different from how Immanuel Kant used it. The agglomeration of primary bodies forms the complex organisms. And similarly, when the division occurs in the larger bodies they might generate the infusoria again. One such example for that would be the generative juices, like semen, of the sexual organs. This is considered the "*Generatio aequivoca*" because it is not the primary creation. The creation can occur only when an organic vesicle forms from inorganic matter. Anything that is larger must have been developed, not created. This includes man as well. And regarding man, Oken supposes that man must probably have developed at a specific spot and at a certain point in time. Further, Oken thinks this supposition is completely in line with the Bible:

> *So the Bible itself teaches us. God did not make man out of nothing; but took an elemental body then existing, an earth-clod or carbon; moulded it into form, thus making use of water; and breathed into it life, namely air, whereby galvanism, or the vital process arose.*

Philosophically, Oken specifies that no individual is eternal. Every single organism is only a specific position in time within the general Galvanism. A specific point in time is finite, but time itself is eternal. Similarly, an individual organism in galvanism is also finite, but the galvanic force is eternal. So when an individual organism is destroyed, a new organism generates. We can consider this as a change in the polarity but not as the loss of the polarity itself. Essentially, the destruction of an individual would mean nothing to nature. Considering that all the larger organisms consist of infusoria mass, they will divide into infusoria when they destroy. This is what putrefaction is. Oken says,

> *Putrefaction is nothing else than a division of organisms into infusoria, a reduction of the higher to the primary, life.*

Oken confirms further that the generation of the primary vesicles (Oken used different terms for these primary vesicles such as Urschleim, protoplasma, sea-mucus, infusoria, and primordial slime) occurs wherever the three vital processes (earth, water, and air) meet. Thus, the generation of organisms from the inorganic matter is one of the important ideas in Oken's work. Though Oken doesn't deal with the notion of heterogenesis, which is the generation of an organism of one species from an organism of an entirely different species, he seems to have no problem accepting it. It is evident when he mentions that the intestinal worms generate from the coagulation of vesicles within the body of the organisms.

One relevant and crucial contribution by Oken is his definition of a very basic form of life called the protoplasma. This protoplasma is a single cell containing nothing but a mucus matter, and except for the fact that it has life it is not much different from the inorganic matter. Though this was an important theory, the fact that it lacked any scientific support limited its influence mainly to the philosophical world.

Karl Marx

After Kant's transcendental idealism, Hegel's dialectical idealism, and Fichte, Schelling, and Oken's nature philosophy, Germany was now ready for materialism. If Kant's transcendental idealism rejected spontaneous generation, one would think that materialism would favor it. And so it did in the case of Karl Marx.

Karl Marx (1818–1883) was a revolutionary economist, philosopher, and socialist. Marx was influenced by Hegel's dialectic method, if not his ideology. In 1844, he met another German socialist Friedrich Engels, and thus began a friendship that continued until death parted them.

Friedrich Engels (1820–1895) was born to wealthy and Protestant parents, but he became an atheist and joined revolutionary groups at a young age. Engels was a socialist by the time he met Karl Marx in 1842 and later again in 1844. Marx moved to London when France, Germany, and Belgium expelled him. Marx lived in dire poverty, and Engels provided financial support for Marx and his family from that point onwards.

Marx and Engels criticized idealism, young-Hegelians group, and even classical materialism for ignoring practical applications to the real world. They worked together in developing Marxist ideology and uniting the working class to form a communist league. In 1848, they published the *Communist Manifesto,* listing the principles of communist philosophy. In their scientific dialectical materialistic ideology, Marx and Engels followed the progress of the great scientific questions of their time. So, the debates on the origin of life came to their attention.

Marx discusses this problem briefly in his 1844 *Economic and Philosophic Manuscripts*. There was no place in socialism for a creation theory involving a supernatural entity. Socialism dealt only with nature and man. Man follows a "circular movement" by producing himself through procreation. And the same goes for nature. So asking about the creation of nature and man is the product of abstraction because this question is based on the assumption that nature is nonexistent. This abstraction has no meaning because we exist and we think, and if nature is nonexistent, then we don't exist. The acceptance of an "alien" being, whose existence is not evident from our "sense experience," to prove the creation of nature and man infers that we also accept the nonexistence of nature, which is not practical.

Marx also refers to the British geologist Charles Lyell's works on Earth's creation and development. One popular theory regarding the Earth's development was Catastrophism, which explained that many sudden short-lived events shaped the Earth's developments. But Lyell supported a different theory, called Uniformitarianism, which suggested that Earth was developed slowly through the processes that continue to work even today. Extending the argument from the creation of Earth to the creation of life on Earth, Marx suggests, spontaneous generation is the only available theory that refutes creation theory:

> *The creation of the earth has received a mighty blow from geognosy – i.e., from the science which presents the for-*

mation of the earth, the development of the earth, as a process, as a self-generation. Generatio aequivoca is the only practical refutation of the theory of creation.

Engels views were in line with Marx's. In 1883, while accepting the recent developments in spontaneous generation theory in his work, *Dialectics of Nature*, Engels also mentions that it is impossible to prove that life did not generate from inorganic matter at any point in time in the history of the world. Life may not generate spontaneously, but it is probable that in nature life may have been generated from non-life matter through a slow, gradual process that probably took thousands of years.

The timing of the *Communist Manifesto* by Marx and Engels couldn't have been more perfect. It was first published on 21 February 1848, and a revolution began in France the very next day. Germany followed suit with the massive demonstrations in Vienna on 13 March. Another prominent German materialist philosopher in the making, Ludwig Buchner, participated in this revolution.

Force and Matter

Luis Buchner or Friedrich Karl Christian Ludwig Buchner (1824–1899) was a revolutionary by nature, like his brother Karl Georg Buchner (1813–1837). Georg Buchner was a talented writer who was greatly influenced by the French Revolution.

After Napoleon's defeat at Waterloo in 1815, which marked the end of the first revolutionary period in France,

Georg formed a secret society based on revolutionary thoughts while studying at the University of Giessen. When it came to the notice of authorities, Georg fled to Strasbourg. He returned to Zurich after two years and worked as a lecturer at the University of Zurich. Unfortunately, four months after he started working, on 19 February 1837 he lost his life to typhus. In a short lifespan of just 23 years, Georg Buchner had etched his name in the history of Germany as a great literary figure.

Ludwig Buchner was only thirteen when his brother Georg Buchner died. In 1848, he participated in the March Revolution, carrying the same spirit as his brother Georg did. But the German Confederation used heavy force to suppress the revolution successfully. Afterward, Buchner joined the University of Tubingen as a lecturer in medicine in 1852. In parallel he was also working on his most impactful work *Kraft und Stoff: Empirischnaturphilosophische Studien* (Force and Matter: Empiricophilosophical Studies). When he got it published in 1855, it created a storm in Germany.

Force and Matter was purely materialistic and atheistic. There was no place for any idealistic, supernatural ideas. Everything, including the mind, came from matter and force that are, in turn, inseparable from each other. For Buchner, everything could be explained scientifically or empirically. The revolutionary tendency of this book appealed to the middle class and working class sections because they could relate it with their own revolution and struggle with the autocracy. But, for the same reason, the book faced strong opposition from certain sections of the society, which cost Buchner his job. Nevertheless, his book saw twenty-one editions and numerous until the beginning of the First World War. And Buchner moved to his native Darmstadt and practiced medicine until the end.

Buchner recognized the implication of vital theory on the origin of life. In his book, he picks a fight with the supporters of vital theory. Taking the example of one of the famous chemists of the time, Liebig, who had rejected spontaneous generation for vitalism, Buchner states that the question of the spontaneous generation is very much an open one. The experiments done in support of it are not superficial as Liebig had claimed, but instead they followed very precise scientific procedures. Buchner remains careful not to exaggerate in these claims, and later points the reader to Liebig's belief in a formative principle in support of the vitalism. But, he says, this formative principle in both organic and inorganic material is internal to them and not, as vitalists claimed, the result of some supreme being acting on our world.

Buchner's goal was not to prove the existence of spontaneous generation. His ultimate goal was materialism, and he strongly advocated for it. For him, the idea of a supreme being compared to any materialistic explanation made less sense. He preferred to accept the unknown as an answer as opposed to any supernatural force.

Even with spontaneous generation, he applied the similar thought process. If we take 'omne vivum ex ovo,' where a living organism comes from its homogenous parents, for granted as a universal law, then, where came the parents from? It can be attributed to either some external power or the generatio equivoca. Buchner stresses that the question of spontaneous generation is far from being settled. There were still many scholars like Dr. Cohn of Breslau, Professor Cienkowsky and M. Pouchet who conducted experiments that seemed to prove the existence of the spontaneous generation. For example, Dr. Cohn of Breslau had observed minute free cells forming in the blood of houseflies that increase rapidly and hence cause the death of those flies.

Buchner did not want to claim something to be true for which there was no clear evidence, so he accepts that the existence of spontaneous generation was not established as a fact. He also points to a limitation of the spontaneous generation – to apply the theory of spontaneous generation for all the organisms that now have the capability to propagate their own offspring seems rather farfetched. The naturalists who believe in a creative power point to this while supporting their belief against the theory. To deal with this problem, Buchner presents an alternative idea that was similar to Anaxagoras's theory to some extent. Per this idea, the germs of living organisms have existed throughout the space for eternity, and they were deposited on the early Earth. These organisms changed gradually over time from lower forms into higher forms. And this explanation would make more sense and less far-fetched than the assumption of a higher power.

Buchner pointed to the theory of evolution in suggesting the gradual development from lower life form into the higher life form. He refers to paleontological investigations that have identified the links between lower organic forms and higher organic forms. Even man is no exception to this law of gradual development. Buchner mentions,

> *Even man, who in his presumption deems himself so*

far superior to the animal world, forms no exception to this law. The Ethiopian race connects him by a number of the most striking similitudes with the animal world. The long arm, the shape of the foot, the thin calves, the long and narrow hands, the flattened nose, the projecting jaws, the depressed receding forehead, the elongated head, the short neck, the narrow pelvis, the pendulous belly, the beardless chin, the color of the skin, the disagreeable odor, the sharp and piercing voice, are all characteristic marks which approach the negro to the ape.

Buchner then gives a complete overview of the origin of life in which spontaneous generation played a pivotal role. In the prehistoric times, Earth was nothing more than a fiery globe that gradually cooled. Organic life started emerging once the temperature reduced and water appeared. Considering water covered most of the Earth, only aquatic animals and plants appeared first. Gradually when more and more land appeared, forests came. Forests helped in clearing carbonic acid from the atmosphere and made way for the land animals. Then came the herbivores, which in turn made way for the carnivores. This is how organic life came into being. In this whole process, the first organisms were the result of the spontaneous generation. Even though we may notice spontaneous generation occurring all the time, it played a pivotal role in the origin of life. Finally, gradual development of lower organic forms into higher forms led to most of the living beings we see around us.

That is how Buchner put forward his theory while also accepting that spontaneous generation was still an open question and gradual evolution of organisms seemed far-fetched. Still, he asserts, his ideas are more plausible when compared to the assumption of a supernatural creative power.

As we can see, spontaneous generation was more of a philosophical need than the scientific one for Buchner. In pure materialism, there is no place for God. If there is no God, spontaneous generation is the best alternative scientific theory that would explain how life originated.

After losing his job due to the strong opposition, Buchner moved to native place Darmstadt. Considering that his book was a sensation and posed a significant threat to idealism, many philosophers opposed to Buchner's book. This led to "back to Kant" movement in Germany in the 1860s. Though it was referred to as "back to Kant," it did not really follow Kant's philosophy. Philosophers preferred a revival of Schopenhauer's version of the Kantian philosophy, which was called Neo-Kantianism.

Even after the rise of Neo-Kantianism, *Force and Matter* continued to be a bestseller. Also, Buchner's materialism became one of the strong foundations for a freethinker's movement. Buchner formed the German Freethinkers League in 1881 in Frankfurt, which became the first ever organization meant for materialists, atheists, and agnostics in Germany.

Evolution

If philosophical needs forced the scholars to form arguments supporting or refuting the theory of spontaneous generation, one major scientific theory needed the support of spontaneous generation: the theory of evolution. The idea of evolution is ancient. Generally, it postulates that all the species are related and the life started with one, or a few, species that diverged gradually to form all the other species. Different theories suggest different mechanisms for the evolution. But there is one critical question that every theory of evolution faces: How did the very first organism come into being? Whoever tried to answer this issue following nature's laws could come up with only one solution—spontaneous generation. For them, crediting the creator for the generation of this first organism seemed to be a far-fetched idea when compared to the spontaneous generation. If evolution is true, as most theorists believed, then spontaneous generation must also be true.

Most of the evolution theorists were not atheists, but since most were also naturalists, they chose to believe in a natural explanation for the generation of organisms and saw no contradiction between their religious faith and these theories. If it was God's will that life must generate from inorganic matter, then who are we to question? But those who took the literal meaning of the religious texts opposed it. Another section's concern was, as Voltaire's was, evolutionary theory combined with spontaneous generation removes God's role completely. This is just enough to push one towards atheism. Moreover, evolutionary theory removes the special place enjoyed by the man among all the life forms. If evolutionary theory is true, then the idea of God creating man with special interest and then creating everything else to serve man would be meaningless. Man would lose that sense of special purpose he had found. And finally, the thought of man evolving from the apes or fishes seemed just absurd.

Until this point, many theories were proposed in support of evolution. But all were rejected because they were based solely on some hypothesis with no evidence or no scientific explanation to support them. A French man, Jean-Baptiste Lamarck, made such an attempt to explain the mechanics of the evolution for the first time in 1802.

Lamarckism

Jean-Baptiste Lamarck (1744–1829) led a very interesting but tragic life. Lamarck was born to a poor family. Following his family tradition, he joined the army but had to retire early due to injuries from an accident. He then worked in a bank office to support himself. Meanwhile, Lamarck read a book by James Chomel and developed an interest in botany. So he studied under Bernard de Jussieu for about ten years. He published the book *Flore françoise* in 1778 on his observations of French flora, which forced the scholars of the time to take note of his name. Comte de Buffon helped him to get a membership of the French Academie des Sciences and also a commission to visit foreign places for studies in botany. By 1792, Lamarck became a professor of natural history, chairing the branch of insects and worms. He was 48 years old.

Lamarck is credited with coining the term "biology" in its modern sense. He also coined the word "invertebrate" and identified ten different groups under the domain. His most famous work is Lamarckism, or the "use and disuse theory," which suggests that an organism would acquire the characteristics over its lifetime and pass them on to its offspring.

Initially a disbeliever of evolution, Lamarck saw how evolution might have occurred and how one species could be related to the other while studying invertebrates. He published his most successful work on the invertebrates in 1801, to which he added a detailed classification of the invertebrates but left out evolution. About a year later he published two more works, *Hydrogéologie* on steady-state geology and *Recherches sur l'Organisation des Corps Vivants* in support of evolution. Both these works were criticized by scholars, as many academics had grown wary of Lamarck's works.

One of the biggest reasons for such criticism and caution was Lamarck's futile attempt at proving a great chemist of the time, Lavoisier, wrong.

Lavoisier

Lamarck went to great lengths defending his theories. In his efforts to do so, he started despising the works of Lavoisier. Antoine-Laurent de Lavoisier (1743–1794), a French chemist, is considered the greatest among the founders of modern chemistry. He rejected the theory that a supposed element called phlogiston residing in combustible matter causes fire. He also refined the chemical nomenclature to a great extent. But unfortunately, Lavoisier found himself in the middle of the decade-long French revolution that began in 1789, which had significant social and political impact.

Lavoisier involved himself in government in 1769 and tried to improve the taxation system from within, and thereby help the peasants. Lavoisier became a member of the Gunpowder Commission at 32 and contributed in improving the quality of the gunpowder. But, as Voltaire had noticed, there was a lot of unhappiness among the general public against the French monarchy, clergy, and a contract-based tax collecting company called the *Ferme générale*. The majority of the public was living in penurious conditions while the members of these three groups had become very wealthy. Such political and social conditions in France triggered the revolution in 1789. While it overthrew feudalism and paved the way for significant improvements in social conditions, it proved to be unlucky for Lavoisier. The Reign of Terror that began in 1793 saw mass executions that claimed tens of thousands of lives. Because of his government work, Lavoisier became a casualty. He was sentenced to be guillotined on the 8 May 1794. Apparently, when an appeal was made to spare his life considering his contributions as a chemist, the judge replied, "The Republic needs neither scientists nor chemists; the course of justice cannot be delayed."

We cannot deny that science benefitted greatly from the revolution, but we cannot also deny that the loss of Lavoisier was the most unfortunate one. A celebrated French mathematician of the time, Joseph-Louis Lagrange (1736–1813) said of the death of Lavoisier, "It took them only an instant to cut off his head, and one hundred years might not suffice to reproduce its like."

When the majority of the chemists accepted Lavoisier's nomenclature and the system of basic elements and followed his standards, Lamarck stood against it. He chose the more traditional four basic elements theory and thought that the scientists who opposed him were too focused on the unimportant small facts rather than understanding the larger picture.

Lamarck tried his hand at many fields like natural history, biology, geology, chemistry, physics, and even philosophy. But his success came only in biology. Though his opposition to Lavoisier's works was related only to chemistry, it stained his reputation in all scientific fields. Many scholars grew increasingly cautious of Lamarck's views.

Steady State Theory

When Lamarck faced the question of the origin of the first life form in his evolutionary chain of species, he chose to explain it using spontaneous generation. He proposed that the simplest living organisms are generated spontaneously through the imbuing of life force on the inorganic matter. These organisms evolved further leading to different complex life forms. But, if all the simple living organisms have already evolved, then why do we still see them? Lamarck's explanation was that the process of spontaneous generation is not a one-time act. It continues to generate simple organisms all the time, and those organisms keep evolving. This "steady-state" theory of biology explained why all the life forms are still seen, along with supporting the theory of evolution. But regarding the scenario in which one form of life comes from the different form of life, Lamark's views are not clear, since his theory of evolution did not require it.

Lamarck was undoubtedly the first evolutionist who supplied evidence to back up his logical assertions. He also emphasized the role of the environment in bringing change, and the capability of the organism to acquire those changes and pass them on to their offspring. But his evidence was not sufficient to convince scholars such as Jean Léopold Nicolas Frédéric Cuvier (1769–1832). Cuvier, a contemporary

French naturalist who had already established himself as a great naturalist and paleontologist, opposed Lamarck's theory aggressively. He was also greatly influenced by Kant's philosophy and, like Kant, opposed evolution as well as spontaneous generation.

In the last decade of his life, Lamarck faced the loss of vision and remained blind until his death on the 18 December 1829. Lamarck was born poor and died impoverished. There was no money to give him a proper burial, so he was buried temporarily in a lime-pit that was exhumed after every few years to make a place for new bodies.

Cuvier wrote an elegy for Lamarck after his demise. In this elegy, Cuvier either ridiculed Lamarck's almost every work except for the one on invertebrates. This indicates how much Cuvier despised the ideas of Lamarck, even if he had admiration for the man.

During the last decade of his life, Lamarck's two daughters took great care of him by sacrificing their own life's time for so long. In his elegy of Lamarck, Cuvier appreciates Lamarck's eldest daughter for her sacrifice:

> *...Her sacrifices, indeed, were carried to a degree which it is impossible to express; when the father could no longer leave his room, the daughter never once left the house. When she afterwards did so for the first time, she was incommoded by the free air, the use of which had been so long unfamiliar to her. It is rare to see virtue carried to such a degree, and it is not less so to inspire it to that degree; and it is adding to the praise of M. de Lamarck to recount what his children did for him.*

Sadly, when Lamarck died, his daughters ended up in extreme poverty, one son was deaf, and another was insane. Only one son, named Auguste, became an engineer and financially stable. Such was the ending of one of the great minds in science.

CHAPTER 4

THE DECISIVE BATTLE

Supporters of the theory of spontaneous generation rejected Spallanzani's experiment disproving the vegetative force for two primary reasons: the prolonged heat killed the vital force needed, and air is a critical factor needed for generation to occur. Here the term "air" was used in a more generic sense because oxygen as we know it today had not yet been discovered. The common usage for the term 'air' was limited to 'clean air' or 'bad air.' But soon after Spallanzani's experiment, many discoveries were made related to the air and oxygen.

Joseph Priestly (1733–1804), an English chemist and theologian, published *Experiments and Observations on Air* in 1776, which dealt with his discovery of the "new" or "purer" air. French chemist Lavoisier later coined the term "oxygen" for it.

Lavoisier had begun his work on combustion theory several years earlier, and all his work related to the oxygen revolved around it. Until then it was believed that any substance that is capable of burning contains an element called "phlogiston" within it. Such substances release the phlogiston into the air when they burn, and once all the phlogiston releases into the air or if the air reaches its limit of absorbing the phlogiston, then the fire subsides.

Lavoisier published a work in 1778 that cited oxygen as the acidifying principle, and in 1783 he attacked the phlogiston theory outright in the paper 'Reflections on Phlogiston'. Unfortunately, Priestly had thought that his "new" air supported the phlogiston theory, and he continued to believe so even after Lavoisier published the experiments suggesting otherwise. Thus Priestly became one of the last significant scholars who defended the phlogiston theory.

These developments had important implications for spontaneous generation as the air was considered a critical factor for the fermentation, putrefaction and the spread of diseases. Joseph Louis Gay-Lussac (1778–1850), a French chemist, conducted experiments that showed that a single bubble of air was enough to bring about fermentation, which, otherwise would not occur. Similarly, a German chemist named Justus von Liebig (1803-1873) came up with a theory that the contact of the wound with oxygen is the cause of sepsis. Both these theories supported the scholars who had refuted Spallanzani's experiment saying there can be no spontaneous generation without air. But before we delve further into experiments related to these theories, it is worth exploring interesting developments in the history of sepsis.

History of Sepsis

In the 4th century BC, Hippocrates introduced the term "sepsis." Two centuries later, Galen noticed that pus formed around a wound in the process of healing. Unfortunately, latter physicians misunderstood Galen's writings, and they started to keep the pus in the wounds thinking it was part of the healing process, and removing it could spread the sepsis. It became a practice to make efforts to generate the pus in the wounds.

Hugh of Lucca, a famed Italian surgeon in the 13th century who practiced medicine in Bologna, refused to follow the tradition of keeping the pus. He condemned the practice and instead advocated to follow some procedures including the cleaning of the wound, the usage of wine as a dressing, and the use of narcotic sponges during surgeries. Though he never documented the procedures, his son and student Theodoric Borgognoni, or Theodoric of Lucca, (1205–1298) documented them in detail. Theodoric improved Hugh's procedures and advocated them. The use of narcotic sponges makes him one of the first users of anesthetics. Theodoric's student Henri of Mondeville (1260–1316) followed his teacher's footsteps and applied the antiseptic procedures successfully.

Unfortunately, Guy de Chauliac (1300–1368), a reputed French physician and surgeon, rejected the antiseptic methods advocated by Theodoric. Instead, he followed more traditional procedures. Chauliac had refused to flee from the city when the Black Death arrived and instead treated the plague patients. His seven-volume treatise *Chirurgia Magna* (Great work on Surgery) on surgery became a huge success and a popular reference text. In this work, Chauliac held to the belief that the pus is beneficial in the healing process, and hence the same was trusted by the future generations at least until the 16th century. In fact, it was only in 1836 that Alfred

Donne, a French bacteriologist, studied pus cells in detail and questioned their benefits.

Many different practices or theories were aimed at stopping the sepsis after the 16th century. Herman Boerhaave (1668–1738), a physician as well as a great teacher, had attributed sepsis to the toxic substances in the air. Pierre-François Percy, a French military surgeon, wrote about his experience in 1785 related to the role of water in dressing the wounds. It goes like this: In 1785, several artillerists were injured when one of the guns burst during a trial. The head of the hospital Lombard tended to their wounds with assistance from Percy. But when the rumor of this incident spread, an Alsatian man came forth and claimed that he knew how to transform regular water into something that can heal the wounds. The commissioner of the province, convinced by the person's account, placed all the wounded soldiers under the care of that Alsatian man. Percy and Lombard his superiors were not allowed to witness his method except for just three days. The Alsatian person took river water, mixed some alum powder and murmured holy sayings in a very low voice. Then the wound was washed with that water, which had now become "holy," and then dressed it with lint. After six weeks all the wounds were healed, which wouldn't have been possible in such a short time with the traditional method. Even though the Alsatian person did not let his secret out, Percy and Lombard were not ready to accept that the "holy chant" had any role in curing of the wound. So he and his partner set out to repeat the same method hoping to obtain the results that are closer, if not the same, to what the Alsatian person had shown. They skipped the religious parts and used only pure, cold water to wash the wound on another group of wounded soldiers. After washing the wound, they dressed it with the usual bandages. In six weeks all the wounds were miraculously healed!

By early 19th century, the theories of Gay-Lussac and Liebig had indicated that the contact of air with the wounds causes sepsis or putrefaction. Most ideas or efforts so far were aimed at finding a way to stop sepsis, but only a few had tried to explain the cause of sepsis. Boerhaave, Gay-Lussac, and Liebig had identified the connection of air with the sepsis but failed to explain the exact mechanics of putrefaction process. How does air cause the putrefaction? Why is it responsible? Unless we answer the "how" question, every reply to the "why" question is subject to suspicion.

'Something' in Air

Franz Ferdinand Schulze (1815–1873) came very close to finding the answers to both the questions. Schulze was a German chemist who invested his talent in the applied science trying to find answers to the stressing problems of the time. He took up the putrefaction problem in 1836.

The aim of Schulze's experiment was to allow air to be in contact with the sterile putrescible matter and see if it would cause the putrefaction. But the condition was that the air must be pure because he believed that it was not the air but something in the air that caused the putrefaction. So he took a flask and filled it with some water. Then he added the putrescible matter into it and boiled it for sterilization. Once the flask was sealed, he had to pass the air. But he did not let air into the liquid directly. Instead, he passed the air through the sulfuric acid so that the acid would kill any foreign substances.

He let the filtered air into the flask for days, but no putrefaction happened. After observing this behavior for enough time he concluded that it is indeed not the air, but some other substance in the air that causes the putrefaction.

About a year after Schulze's work was published, a very similar experiment conducted by another German physiologist named Theodor Schwann (1810–1882) came to light. Schwann's work made more news because he ventured farther than Schulze by extending the implications of the outcome of his experiments to another hot topic of the 19th century: fermentation.

Role of Infusoria in Putrefaction and Fermentation

At nineteen, Theodor Schwann met his mentor Johannes Peter Müller (1801–1858) while studying, first at the University of Bonn and again at the University of Berlin. Muller is best known for his masterwork *Handbuch der Physiologie des Menschen* (Elements of Physiology), which marks an important milestone in physiology. When Schwann helped with his experiments, Muller was still working on this book. But Muller was a dedicated believer in the theory of vitalism. Vitalism has its unique place in the history of theories of the origin of life.

Vitalism

We find the traces of vitalism in both eastern and western ancient history including Chinese, Indian, Egypt, and Greek histories. As the name itself indicates, vitalism suggests that there is some vital energy in all the living things that separates them from all the inorganic matter. All living organisms are primarily different from inorganic matter because of this vital energy. This is the essence of vitalism. Many referred the vital energy as the soul itself because of its very nature. Vitalism distinguished living things from the machines by implying some higher purpose for them.

Vitalism gave a boost to the theory of epigenesis. Epigenesis said that living organisms develop gradually from a single egg, but what causes such gradual development? Who or what directs the egg that has no preformed organs to develop into a perfect living organism? Vitalism answered these questions. Caspar Friedrich Wolff (1735 – 1794), a critical figure in the field of embryology, was a supporter of the epigenesis and vitalism. And as we have seen, Needham referred to a vegetative force while explaining the spontaneous generation, and he was also a supporter of epigenesis.

In the early 19th century, vitalism was made popular by a Swedish chemist Jöns Jacob Berzelius (1779–1848). Berzelius is a well-known name in the chemistry world. Berzelius lost his parents at an early age and grew up under the care of a relative's family. He studied at the Uppsala University and worked as a physician for a while before becoming a professor at the Karolinska Institute. By 1818, he was appointed as secretary of the Royal Swedish Academy of Sciences. He coined many key terms like polymer, protein, and catalysis. Berzelius also classified many chemical elements such as silicon and thorium. The chemical notation system for the elements, which is still in use today, was proposed and initiated by him.

Berzelius also coined the phrase "organic chemistry" while distinguishing organic and inorganic compounds for the first time.

Berzelius firmly believed that the vital force was a prerequisite for any life. He further maintained that organic compounds could not be synthesized from inorganic compounds. Organic compounds are any chemical compounds that contain the carbon element and comprise the basic building blocks of living organisms.

In 1828 Berzelius's student Friedrich Wöhler (1800–1882) gave the first, but far from being the last, blow to vitalism when he accidentally synthesized urea, an organic compound, from inorganic reactants in an attempt to synthesize ammonium cyanate. Until then, urea had only been obtained from organic sources. But still, even this breakthrough failed to have any impact on the popularity of vitalism at the time. It took more than a century for the vitalism to become obsolete.

Experiments Related to Putrefaction and Fermentation

Johannes Peter Müller was a faithful believer of vitalism, and that reflects in his magnum opus work 'Elements of Physiology,' which became a standard text in physiology. But Theodor Schwann did not agree with his teacher.

Schwann has some of the very critical discoveries in his name. In his significant work on the digestion in 1836, he discovered and coined the term "pepsin," which is an enzyme that degrades protein into smaller units called peptides. Schwann also discovered the cells, now called "Schwann cells," that protect the peripheral nervous system.

A year after Schulze did his experiments related to the

putrefaction, Schwann came up with the experiments of a similar nature. He placed meat in a closed flask containing water and then boiled it. Once boiling was done, he supplied pure calcined air into it. As he had expected, there was no spontaneous generation in the flask, proving the same point as Schulze. But Schwann went a step further. He wanted to show that life would not be affected in any way when the calcined air is used. So he conducted further experiments on the animals like frogs and confirmed that calcined air would have no adverse impact on the life.

Schwann came to two important conclusions after the experiments:

First, the reason for putrefaction is the growth of infusoria. Until then there were a lot of attempts to stop the sepsis from happening, but nobody had discovered why it occurs. Schwann confirmed that the dead flesh did not putrefy when subjected to calcined air, but when subjected to the normal open air it decayed. But instead of attributing putrefaction to the air, as Gay-Lussac and Liebig had done, Schwann identified the role of infusoria. The infusoria that are already part of the air grow and reproduce at the expense of flesh.

Second, fermentation happens when the living organisms like yeast enter the mixture and reproduce, and not because of the introduction of air into the mixture, as many scholars had suggested. For example, Gay-Lussac had claimed that a single bubble of air was enough to trigger the fermentation. But Schwann had noticed that passing the calcined air did not initiate fermentation, but instead the process began as soon as the yeast appeared.

If, as Schwann suggested, fermentation and putrefaction

processes depended on the living organisms then it would have a significant impact on the veracity of spontaneous generation. How could such small organisms be part of critical chemical and biological functions?

Fermentation - A Chemical Process, or, A Vital Process?

Fermentation is perhaps one of the oldest mechanisms of civilization. There is evidence to show that men knew how to ferment even in the 7th millennia BCE. The fermentation process was ancient, but nobody knew why it occurred. Antonie van Leeuwenhoek did notice yeasts under the microscope, but due to their immobile nature, he could not make out if they are living animalcules. Berzelius suggested that the role of yeast was just a catalytic one, and fermentation process as a whole was purely chemical. Unsurprisingly, this supposition was perfectly in line with his vitalism theory. Along with Berzelius, the other influential chemists like Lavoisier and Gay-Lussac viewed fermentation as a purely chemical process. As per them, yeast had no significant role other than some catalytic effect in triggering the reaction. Moreover, yeast was not even considered a living organism. Given this background, Schwann's discovery becomes a critical one with results being contrary to the widely held belief. Schwann even observed the reproduction of yeast, confirming that they are indeed living organisms.

A French physicist named Charles Cagniard de la Tour who was working independently on the fermentation subject also came to the same conclusion that yeast is a living organism. With support from these two independent experiments, one would expect that it would have been easier to convince other scholars. But it did not happen.

Most of the famous chemists panned Schwann as well as La Tour's results. Berzelius held that the fermentation is still a chemical process and remained reluctant to accept yeast as a living organism. Liebig, who had mentioned that contact of the wound with air would cause sepsis, supported Gay-Lussac firmly and stated that fermentation is indeed a chemical process. Liebig pointed to certain type of fermentations, like lactic acid formation, which does not require yeast at all. He also added that even if the yeast were living it wouldn't make much difference because it only acts as a catalyst. When yeast disintegrates, its substance released into the fermentation matter must be causing the catalysis and hence increasing the rate of fermentation. With all such mammoths in the chemical field backing the purely chemical nature of the fermentation process, Schwann and La Tour's findings were suppressed.

Particle Filter Experiments

In 1843, a new person in the field, Hermann von Helmholtz, questioned Liebig's theory of decomposed yeast causing chemical catalysis. Helmholtz used a thin membrane to separate a fermenting liquid from the fermentable liquid that was not yet fermenting and lacked any yeast. He noticed that even though liquid did exchange between both the sides through the membrane, no fermentation took place on the side where it was not already happening. This proved that there was "something" in the fermenting liquid that was not soluble and that caused the fermentation. It must have been a big enough solid particle to fail to pass through the membrane, which suggested the yeast. But this experimental evidence was also suppressed. Liebig refused to budge from his stand on the matter.

After Helmholtz, two more experimenters, Heinrich Georg Friedrich Schroder (1810–1885) and Theodor von Dusch (1824–1890) came up with a simple but effective method of filtering the air particles by using cotton to preserve the food. They noticed that by using cotton wool one could prevent living particles from entering any food or liquid. Schroder even successfully repeated Helmholtz's experiment. But still, the establishment dug in its heels.

Enter Pasteur

Pasteur, Louis, 1822-1895

Pasteur (1822–1895) was born in the Jura District of France. His father Jean-Joseph Pasteur owned a small tannery that was their only source of income. Louis Pasteur was not a bright student. His first attempt to enter the ENS failed, but in 1844 he managed to make the entrance test the second time. At ENS, Pasteur found a mentor, Antoine Jerome Balard (1802–1876), who allowed him to use his lab during free time.

Balard came from a very poor family, and his godmother helped with his studies. At the age of 22, still a student, he discovered bromine, and in next two years his treatise was confirmed and accepted by the Academie des Sciences in Paris. After graduation, he joined the pharmacy school as a teacher in Montpellier, his hometown. Later he moved to Paris, first working at Sorbonne and then Ecole Normale where he would stay until 1851.

Being a supportive mentor, Balard encouraged Pasteur at Ecole Normale. But after graduation, Pasteur was appointed as a physics professor at Collège de Tournon (now called Lycée Gabriel-Faure) at Ardèche and had to move. Both Balard and

Pasteur did not like this new situation, so before long Pasteur was back at the ENS as an assistant to Balard. During this period, Pasteur worked extensively on crystallography.

Molecular Chirality

Tartaric acid is a crystalline acid that is found naturally, such as in grapes. After the fermentation of wine, it was normally found in the residue, and Pasteur took the task of studying its properties. He noticed that tartaric acid caused the rotation of the plane of polarization of light. But what was more surprising that the tartaric acid produced through chemical synthesis did not cause the optical rotation. Upon closer observation, he found that the chemically synthesized tartaric acid was not exactly tartaric acid in the real sense. Chemically synthesized tartaric acid had two crystals in it, one of them exactly like tartaric acid, and another was its mirror image. For example, the left and right hands of a person make mirror images of each other. If we rotate our right hand in whatever angle possible, it will not result in the exact image of a left hand, but it would still look similar to the left hand. Pasteur noticed this non-superimposable mirror image structure and managed to separate both the crystals and test the optical activity. One crystal caused the plane of polarization of light to turn right while the other turned it in the opposite direction. Pasteur had just discovered molecular chirality, as it was coined by Lord Kelvin in 1894. Natural tartaric acid, or (+)-tartaric acid or "right" tartaric acid, and its mirror image, or (-)-tartaric acid or "left" tartaric acid, were optically active, but the chemically synthesized "racemic" acid, or (±)-tartaric acid, was not. The optical activity was not new when Pasteur made this observation. It was noticed and confirmed in the early 19th century, but Pasteur was the first to observe and confirm the cause.

This first major work by Pasteur became the strong base for his future adventures. Balard was impressed with Pasteur's work and made successful attempt to bring him to the notice of the members of Academie des Sciences. Balard's mention of Pasteur's experiment aroused great interest in one member, Jean-Baptiste Biot (1774–1862). Biot had made a lot of contributions in the field of optics, specifically related to the polarization of light, and Biot's works on polarization seemed to support corpuscular or particle theory of light. The corpuscular theory of light was an extension of atomism theory applied to the light, which claimed that light comprised particles. As per the philosophy of atomism, there are two fundamental principals in the universe: atoms and void. Though Pierre Gassendi proposed the corpuscular theory of light in the 17th century, it was Newton who had developed it further. By 1815 Biot had noticed the rotation in the plane of polarization of light passing through the organic liquids and gases.

Biot asked for a demonstration of Pasteur's experiments and monitored it personally. Biot was completely satisfied with the results, and he was so happy that as Pasteur recalled later:

> *...he[Biot] took me[Pasteur] by the arm and said: "my dear boy, I have loved science so much all my life that this touches my very heart."*

Biot had come from a financially stable family. He survived

the political turmoil of the French Revolution, witnessed Lavoisier's execution, and achieved a name for himself in the scientific community. Laplace was his mentor as Balard was for Pasteur. He made contributions in the fields of astronomy, mathematics, magnetism, and optics. Pasteur had now gained the attention and support of a man of strong repute. With Biot at his back, Pasteur became a well-known name at the Academie. Biot became close to Pasteur and corresponded with him frequently. When Louise and Marie Pasteur had a son Jean-Baptiste in 1851, Biot became his godfather.

Fermentation and Microorganisms

Pasteur continued to work on crystallography during next few years of his career. He also worked across different universities like at the Dijon Lycée, University of Strasbourg, and Lille University before moving back to ENS in late 1857.

While appointed at the Lille University, a student's father, Mr. Bigot, approached Pasteur with an interesting problem related to the synthesis of alcohol from beetroot juice. The alcohol produced would turn sour, and the container would smell foul. Bigot sought any possible help with that problem, and Pasteur gladly took charge of the investigation. Until then Pasteur had absolutely no involvement with biology. But he applied his deterministic mind and invested all his time and energy into solving this problem. After work at the University, he would visit Biot's factory to collect the samples and then conduct experiments with no sense of time. As he had said in the opening speech while taking charge as the dean in 1854, "*chance only favors the mind which is prepared.*"

Pasteur saw that when fermentation produced good quality alcohol, only "*plump and budding*" yeast cells were present. But whenever it turned sour there were tiny rod-like microbes along with the yeast cells. The presence of yeast cells in the fermentation process was already a recognized matter, but its exact role was still a debated topic. Schwann, La Tour, and Helmholtz's findings did not gain wide acceptance due to the influence of the likes of Liebig, Gay-Lussac, and Berzelius. Now Pasteur had found himself in the middle of the very same debate.

Pasteur conducted an experiment he was already very comfortable with. He found that the complex chemical compounds were being formed during the fermentation process and subjected them to the optical activity experiment. He saw that the plane of polarization of light changed, which convinced him that fermentation is very much a vital process. It was a common understanding at the time that optical rotation is caused only by the naturally obtained organic compounds. So not surprisingly, Pasteur the optical rotation to be an indication of the molecules produced only by the living organisms.

His observations led Pasteur to suspect that the rod-like microbes are the cause of sourness. He confirmed that these rod-like microbes produced the lactic acid, which contaminated the alcoholic liquid. The lactic fermentation along with the alcoholic fermentation causes the sour taste. Pasteur presented his paper on the lactic fermentation in 1857, explaining the role of rod-shaped microbes in the lactic acid formation.

After his work on the lactic fermentation, Pasteur conducted further experiments related to alcoholic fermentation.

He took a fermentation medium that included glucose and ammonium salts and added yeast to it. Yeast would grow with no further interference, and fermentation would also take place. But if he kept the yeast sterile then they would not grow, and even fermentation would not take place. Passing the air made no difference. This was a counterattack to Liebig's argument that chemicals from the decaying yeast contribute to the fermentation process. Now Pasteur had proved that unless there is a biological growth of the yeast cells, there would be no fermentation.

Near the end of the same year, 1857, Pasteur moved back to ENS and took charge as the administrator and director of scientific studies. On 21 December 1857, he presented his paper before the Academie on alcoholic fermentation, along with the results of his experiments. Pasteur concluded that fermentation is a biological process and different types of microbes cause different types of fermentations. He also gave a simple solution for obtaining the left tartaric acid. By subjecting paratartaric or racemic acid to a certain fermentation method, he had found that only right tartaric acid undergoes the fermentation but the left tartaric acid doesn't. So at the end of the fermentation process, only the left tartaric acid will remain.

Liebig did not agree with Pasteur on his findings. Until his death, he held to the belief that fermentation was a chemical process. Both Liebig and Pasteur were proved partially correct only in the in the late 19th century with the discovery of the role of enzymes in yeast. Nevertheless, Pasteur's findings received wide acceptance due to the practical outcomes in the wine industry. Pasteur continued to work with microbes to come up with solutions for the pressing problems of the time.

Felix Pouchet

Félix Archimède Pouchet was one of the people who criticized Pasteur shortly after his presentation on the fermentation. Pouchet (1800–1872) was a French naturalist and an avid experimenter. A year after completing his medical degree in 1827, he became the director of the Rouen Museum of Natural History and its botanical garden. Pouchet was also awarded the Order of Knight of the Legion of Honor in 1843 . He had also secured the membership of the Academie des Sciences, while Pasteur had to wait until 1862. Pouchet was better known in scientific circles on account of his experience, but Pasteur had the advantage of being part of the close circle of senior Academie members. Pouchet also happened to be a staunch supporter of the theory of spontaneous generation, and that led to his conflict with Pasteur. The idea that the growth of yeast and other infusoria caused the fermentation did comport with the concept of spontaneous generation, which would have required the act of fermentation itself to generate the microbes.

In 1858, Pouchet came up with series of presentations before the Academie supporting the spontaneous generation. For him, spontaneous generation was a very scientific process. So his experiments aimed at finding the exact scientific methods resulting in different life forms. And he devised experiments to determine just that. As a result, he published a major work entitled *Heterogenie* in 1859.

For Pouchet, "heterogenesis" meant the spontaneous generation from specific ingredients that included vital force, organic matter, air, water, and temperature. All these ingredients must meet certain conditions; too high a temperature or too low a temperature than what is needed will not result in the spontaneous generation. He then conducted many putrefaction and fermentation experiments to examine and verify the same. For Pouchet's heterogenesis, organic matter was an essential ingredient for the spontaneous generation to occur. Pure inorganic matter alone generating the organic being, called the "generatio aequivoca," was not his domain.

The sharp criticism of his work on fermentation made Pasteur notice Pouchet . When he thought about working on the problem of spontaneous generation, his well-wishers Biot and Jean-Baptiste Dumas (1800-1884) told Pasteur it would be a waste of his valuable time because the problem of spontaneous generation is such that one can obtain no fruitful outcome from it. Dumas himself had criticized Pouchet's work on spontaneous generation to some extent, but he could not refute the experiments completely with no evidence at hand. But Pouchet's presentations, the publication of the book, and his criticism of Pasteur's fermentation work drove Pasteur to prove him wrong. And so, Pasteur took the thousand-year-old question upon himself to answer and put an end to it once and for all.

The Battle

When the conflict between Pasteur and Pouchet escalated, and the question of spontaneous generation was once again a hot topic, the Academie announced a prize of 2500 francs for the person who would come up with experiments determining the existence of spontaneous generation. While Pasteur had to work hard to devise an experiment method in refuting the theory, Pouchet didn't have to put much effort as he had already published a famous work.

Pouchet's Experiments

Pouchet had half-filled a flask with the boiling water and hermetically sealed it. He then inverted it carefully over a small container filled with the liquid mercury. Next, he removed the seal from below the mercury surface and passed calcined air. This ensured that the two vital components, air, and water, required for the spontaneous generation were brought together with no contamination from the atmosphere. All that was needed was some organic matter. So Pouchet took about ten grams of hay from a sealed vial, which had been placed in an oven for about thirty minutes at 100^0 C, and carefully placed the hay inside the flask from below the surface of liquid mercury. Now, all the ingredients of life were placed together with utmost care taken in maintaining a pure environment. Then he waited, patiently. And then, after eight days of long wait, there it was! Mold formed on that mixture, and Pouchet deemed the experiment a success.

This experiment, per Pouchet, proved that the atmospheric germs couldn't be causing spontaneous generation. He went a step further by claiming there are no germs in the atmosphere at all. And no infusoria in the atmosphere means they cannot contaminate the infusion and thus cannot be a factor in spontaneous generation.

Pouchet also alluded to Gay-Lussac's experiment showing that a single bubble of oxygen brought forth fermentation. Different experimenters had repeated Gay-Lussac's experiment successfully using different infusions. Different fermentation matter must require different infusoria, and if that is so, does this imply that a single bubble contains many types of infusoria within it? To Pouchet, if a single bubble of air contained so many kinds of infusoria that each organic matter could pick a germ specifically suited for it, then it would form a dense fog in the atmosphere.

Atmospheric Germ Theory vs. Spontaneous Generation of Germs

Pasteur went through Pouchet's line of argument and understood that he had to figure out exactly where Pouchet's experiment went wrong. And the most important challenge ahead was to prove the existence of infusoria or germs in the atmosphere.

Pasteur decided to first investigate Pouchet's experiment, so he meticulously followed Pouchet's procedures. In Pouchet's experiment, atmospheric germs must have contaminated the organic mixture at some point, but the question was where and how. The answer must lie in one of the ingredients. The air was pure because it was chemically synthesized. Hay taken from a 100-degree oven would have no infusoria. Water boiled for 30 minutes could have no contamination. The flask used was also sealed shut until it was carefully placed under the surface of mercury, so even that could not have been contaminated. Only another item used in the experiment was mercury, so Pasteur suspected that contamination must have occurred there. Pasteur took notice of the dust particles in the mercury and compared them with the infusoria. We can observe the dust particles in the atmosphere when a ray of light passes through a considerably darker room. These particles float in the air but eventually, very slowly, end up on the surface of the objects in the room. Here, mercury would be that object. From the time extracted from the mine, till the time used for the experiments, mercury must have been exposed to a lot of dust particles. And when we place anything, like the flask, in the mercury these particles make their way inside. One can easily apply this principle of dust to the atmospheric germs. Along with the dust, germs too got inside the flask, thus contaminating the mixture.

To test this theory, Pasteur conducted a simple test that was similar to Schroder's cotton-wool experiment. Take a glass tube with an open end at both the sides. Then place a cotton wad at one end of the tube and let the water pass through it. After a while, we can visibly notice the dust particles attached to the cotton wad. Then take a small dish with some water in it and squeeze the cotton wad. Once the dust particles move into the water from the cotton, we can subject that water to the microscope observation and clearly see the presence of microorganisms. Some of the microorganisms Pasteur had found using this method were similar to what Pouchet discovered in his mixture, which proves those germs were not produced spontaneously.

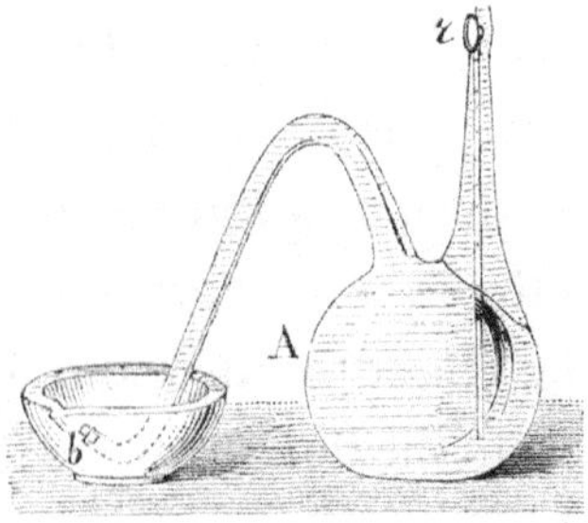

Pasteur's flask

So Pasteur first took the organic matter infusion into a long neck flask. This flask's neck was thin and long, but it was not bent. When he boiled the infusion to rid it of all the germs and let it cool, infusoria still appeared within a few days. This happened because the flask was open and its neck, though long, was straight, so germs could still enter through simple gravity. He then repeated this process using a swan-neck flask. This time, even though the flask remained open, infusoria did not appear no matter for how many months or years he waited. As Pasteur claimed, each flask remained open to the air since the time he conducted that experiment in 1860 until his presentation before the committee in 1862. But, while air could pass through that thin bent neck easily the solid particles could not. So the dust and the infusoria that were part of the air entered the neck but never reached the infusion. Now, if one wanted to confirm that was the case then all he had to do was to shake the flask well, so the incoming air pressure will move all the dust and the infusoria into the flask contaminating the infusion. And a few days later, he would notice mold just like Pouchet did.

Pasteur wanted to find an answer for one more problem

- Does infusoria exist everywhere in same numbers? As Pouchet had pointed out, a huge number of infusoria in a single bubble of air would mean a thick fog all around us. To find an answer to this, Pasteur undertook an extensive study. He took a set of about twenty swan-neck flasks containing the infusion and boiled them. While boiling, he sealed the flasks by melting the glass at the open end of the neck. Doing this ensured that infusion was free from the germs, and as expected, no infusoria appeared in any of them. After a while, he broke the necks, and the atmospheric air gushed in. Once ensured that air was inside, he sealed the flasks again. But he did not boil the infusion this time so, as the atmospheric air had entered the flasks, one would expect the infusoria to appear after few days. And as expected, the infusoria did appear in the flasks. But the difference was that not all the flasks had the infusoria. Based on the location where Pasteur conducted the experiment, there was a difference in the number of flasks in which infusoria appeared. He repeated this experiment in his laboratory, at his home, in his garden, and even on mountains and glaciers. At the inhabited regions like the home, garden, and laboratory, he found that infusoria appeared in the majority of the flasks. On top of the Jura Mountain, infusoria appeared in only five out of the twenty flasks. He repeated it in a county at the bottom of that mountain and found that eight of twenty flasks had infusoria. Then he went to the glacier at Mont Blanc, and infusoria appeared only in one flask. So Pasteur concluded that the distribution of the germs in the air is not same everywhere. There are more infusoria in the air in the inhabited locations than in the less inhabited regions.

Victory

Pasteur was all set to face Pouchet and his cohorts before the committee. The committee of five members formed by the Academie comprised of a zoologist Henri Milne-Edwards (1800–1885), a botanist Adolphe-Theodore Brongniart (1801–1876), a physiologist Marie Jean Pierre Flourens (1794–1867), a naturalist Jacques Marie Cyprien Victor Coste (1807–1873), and another eminent physiologist Claude Bernard (1813–1878). Pouchet and his partners, zoologist Nicolas Joly (1812–1885) and Joly's student Charles Musset, who had submitted papers supporting heterogenesis, were not happy with the committee selection. Most of the committee members had previously criticized Pouchet's works. But, they had no evidence to prove that spontaneous generation doesn't happen or to establish that Pouchet's experiments were flawed. And two of the committee members, Milne-Edwards and Flourens, were mentored and influenced by Georges Cuvier who vehemently opposed the idea of transmutation, which is nothing but evolution, and discredited Lamarck. Cuvier did not involve himself in the topic of spontaneous generation directly, but being opposed to evolution he supported the idea that every living organism came from its parents. It was no secret that Milne-Edwards and Flourens felt similarly.

Pouchet, Joly, and Musset decided that the committee members, who had been critical of their views and were opposed to the theory of spontaneous generation, were partial to favor Pasteur and thus withdrew their papers from the competition. With Pouchet, Joly, and Musset gone, no other eminent French scientist stepped up to defend spontaneous generation. Gay-Lussac and Berzelius were no more, and Liebig was a German. With no other contenders in the competition, the Academie declared Pasteur winner. But his easy win doesn't make Pasteur's presentation weak. He was one of those rare scientists possessing excellent presentation skills. He wouldn't rest until he had all the answers, and then he would present it to the world with the same passion.

The committee members were satisfied with Pasteur's experiments, but Pouchet, Joly, and Musset were not. Pouchet and Joly criticized Pasteur's presentation in his celebrated memoir *On the Organized Corpuscles That Exist in the Atmosphere* in 1861. Pouchet called the idea of the atmospheric germs seeding the life in the infusion as 'panspermia.' Joly asked if the germs are highly concentrated in some locations and not in others, then why would chemical analysis so far have failed to detect them? Pouchet and Joly repeated Pasteur's experiment by climbing the Glacier Montanvert and claimed to have observed the spontaneous generation in all of their flasks.

Despite his victory, Pasteur was not happy with such criticism. For him, spontaneous generation was no longer a question; its nonexistence was a matter of fact. So he was convinced that Pouchet and Joly's experiments were flawed. His experiment required a careful, clean, and error-free method, and Pouchet's team must have let germs into the flasks somehow. He expressed dissatisfaction towards their criticism in letters to his father, friends, and mentors like Balard. Balard assured him that truth was on his side. But when Pouchet and Joly demanded another demonstration of

Pasteur's experiments, the Academie happily agreed to it and set a date in 1864.

Pouchet and Joly were once again upset with the selection of the committee members. This time the committee featured Balard and Dumas along with Milne-Edwards, Flourens, and Brongniart from the previous committee. As everybody knew, Balard and Dumas were mentors and friends of Pasteur. Balard had a big role in shaping Pasteur's career and had also personally suggested the idea of the swan-neck flask. Dumas too favored Pasteur. He even played a prominent role in promoting Pasteur's name for the Academie's membership. So once again Pouchet and Joly withdrew from the challenge, doubting the integrity of the committee members.

While Pouchet and Joly continued to attack Pasteur's experiments in public, Academie concluded its position on spontaneous generation. Pasteur's presentation was effective and powerful. His experiments were precise and flawless. Academie credited Pasteur for putting the question of spontaneous generation to rest once and for all. The debate and the challenge between Pasteur and Pouchet had made big news in public, and clearly, Pasteur had upper hand.

Preconceived Notions

Every theory starts with a simple assumption, and a preconceived notion influences every assumption. So when a scientist undertakes a problem to solve, he begins with an idea. In some cases, like with spontaneous generation, this preconceived notion tends to be very strong. Based on this notion he forms a hypothesis to prove it. But the outcome of the experiments may contradict his personal views. He may further move in the wrong direction if he fails to notice any possible flaws in the method of experiment. The chances of this happening were high in case of Pasteur and Pouchet because the field they were working on was still in a developmental stage.

Both Pouchet and Pasteur were religious. But the way they interpreted how spontaneous generation relates to their beliefs differed, hence their preconceived notions also differed. Pouchet was a firm believer in heterogenesis. He never ventured to say that the creation of first life was purely materialistic, as Buchner did, with no divine intervention. He thought the first creation was a special one-time spontaneous generation triggered by a divine action. After that, spontaneous generation continues to happen all the time. Pouchet saw God's hand in the heterogenesis process because a vital force was necessary for it to occur.

Buchner helped start a materialistic or atheistic movement in Germany, and spontaneous generation was one of the subjects he used to support his materialistic philosophy. He had even made references to Pouchet's experiments. On the other hand, many in the church considered the idea of spontaneous generation as a threat to the idea of divine creation. Being an ardent Catholic, even Pasteur felt the same. His stand on this subject echoed the larger sentiment. Pasteur expressed concern over the dangerous implications of spontaneous generation in an impressive speech at the Sorbonne Scientific Soiree on 7 April 1864. But, at the same time, he stressed that the religious or philosophical implications have nothing to say when it comes to science because all one must do is follow the science alone without thinking about anything else. So, even though Pasteur had a strong scientific mindset, we cannot deny that he had strong preconceived notions against spontaneous generation due to his religious beliefs.

Perhaps religious and philosophical reasonings motivated both Pouchet and Pasteur's opinions. But one cannot accuse them of tampering with the results of their experiments, even though Pasteur chose to cherry-pick only the successful trials, as his notes later showed. For Pasteur, the negative result indicated that either he had made some error or was not careful enough while conducting the trials, so he simply separated and ignored such results. It never occurred to him that the steps followed in the experiments had been perfect but still resulted in negative outcome. But also, cherry picking the results was not considered an ethical issue when this incident happened Pasteur was one of the most disciplined scientific persons in his era, and it is highly unlikely that he would resort to any unethical action.

Even Pouchet, Joly, and Musset were experienced scientists. Even though they were criticized for their experimentation procedures once they lost the battle with Pasteur, Pouchet had a strong reputation for conducting experiments prior to this. And the trio faithfully repeated the experiments done by others like Schwann but came up with a different result. They repeated Pasteur's experiment with same care, but again came up with opposite results. Interestingly, as we will see in next chapter, even if Pasteur followed all the steps perfectly, it was still possible to observe spontaneous generation.

Germ Theory of Diseases

Pouchet continued to reject Pasteur's findings, and three years before his death he published an encyclopedia called *The Uni-*

verse, in which he ridiculed Pasteur's work. But Joly, witnessing the success of Pasteur's works, conceded defeat: "*If our discussions contributed Pasteur in getting such wonderful results in any way then we will rejoice from the heart, and we will forget that he was a bitter and discourteous opponent.*"

Pasteur not only asserted that germs are not generated spontaneously, but also that they exist all around us in the atmosphere. He also attributed them as the cause of diseases. Pasteur's works laid a strong foundation for the germ theory of diseases, but in the past, impressive results were brought forth by different researchers with similar proposals. But all of them either failed to understand that germs in the atmosphere are the cause of diseases or failed to apply the results obtained to practical medicine.

Germ theory of diseases is a much larger subject than this book can properly address, but it is worth examining two smaller yet related topics. First one is the theory of humorism. Humorism was the most widely accepted disease theory from ancient times until the 17th century. And the second is a Hungarian physician named Semmelweis. Semmelweis's works came just before Pasteur's decision to investigate the spontaneous generation. The failure of Semmelweis in convincing the world about the existence of invisible contagions that cause the deadly diseases clearly shows us how close other investigators were in coming up with the same theory that Pasteur did.

Humorism & Miasma

Humorism is a system of medicine that attributes the

cause of the disease to an imbalance of bodily fluids called humors. There are four humors: yellow bile, black bile, blood, and phlegm. These primary humors were linked with the basic elements: air, water, earth, and fire. This system of medicine concentrated its efforts on identifying the cause of excess or deficit humors in the body. This method of disease identification and treatment was followed by most of the major ancient medicine systems and continued for more than two millennia afterward. Old schools of medicine that are still alive today, like Indian Ayurveda, use this theory of disease. Hippocrates used the humor system in ancient Greek history, and Galen later popularized it. It dominated the medical field until the 19th century. The major blow came when Rudolph Virchow published *Cellular Pathology* in 1858 in which he extended Schwann's cell theory to add a third dictum: all cells come from pre-existing cells. Virchow's work contradicted humorism by considering disease at the cellular level and even by comparing the diseases between humans and animals.

If humorism was the answer to "what caused the disease?" then miasma was the answer to "how did it transmit?" When a body part decomposes, the particles called "miasmata" emanate from it to spread through the environmental mediums like air and water. A bad smell can identify the presence of miasmata, and people who come in contact with it could get the disease. This theory gained wide popularity, especially after the Black Death.

Semmelweis - Savior of Mothers

Hungarian physician Ignaz Philipp Semmelweis

(1818-1865) joined Vienna General Hospital in 1846 as an assistant to Professor Johann Klein in the obstetrical clinic. There he noticed a strange pattern in deaths due to childbed fever. In general, there were three childbirth locations at the hospital. Two of them were clinics, and third was street birth for women who wouldn't or couldn't reach the hospital on time. Semmelweis noticed that the first clinic had much higher mortality rate than the second clinic. The fact that even the street births had a lower mortality rate when compared to the first clinic was even more puzzling. Semmelweis tried his best to isolate the cause, and he finally concluded that only possible difference was the people involved. The first clinic was for the medical students while the midwives handled the second clinic. But that did not explain the reason behind the difference.

Haunted by the deaths in the maternity clinic, Semmelweis left for Venice on 2 March 1847 to take a break. When he was out, Semmelweis's colleague and friend Jakob Kolletschka (1803–1847), a forensic professor at the hospital, was performing an autopsy one day with the help of some of his students, and got his finger pricked by a knife. Because of this Jakob fell ill, and a few days later he passed away. Semmelweis returned a week later on the 20th March after hearing the bad news. Semmelweis admired Jakob, so his death had affected him greatly. But he noticed that the symptoms of Jakob's illness were exactly same as the women who died because of the childbirth fever. Semmelweis performed the autopsy on Jakob's body and confirmed his observations. The connection between the autopsies, contamination, childbed fever, and clinic's high mortality rates struck him all at once.

Autopsies were performed only in the first clinic, and the students would come directly to the maternity hall from the autopsy rooms. As the humorism theory still dominated at the time, there were no procedures in place to eliminate the potential contamination. In the second clinic, midwives would never perform autopsies, so there was no chance of contamination.

Semmelweis was not aware of the possibility of a germ contamination, but he knew that there was *something* contagious that was carried from the autopsies to the maternity clinic. He did not think of devising some experiment to prove his idea, and he did not put more effort in finding the exact nature of the contagious matter. Convinced by his own theory of contagion, Semmelweis set forth to implement a solution to prevent any more tragic deaths in his clinic.

Semmelweis had noticed that using chlorinated lime would remove the bad smell from the autopsy organs, and he hoped that using it to wash the hands of physicians would eliminate whatever contagious matter one carried with him after the autopsy. Semmelweis enforced the policy, and he quickly noticed that the mortality rate drastically reduced from 18% in April to 2% by June. And in next few months, this reduced even further. Semmelweis was happy that he had found a solution to one of the most pressing matters of his time, but his happiness did not last long.

Semmelweis had to face a lot of criticism from many well-known authorities in his field. Except for managing to bring down the mortality rate, there was nothing else to support his theory. Semmelweis knew nothing about the nature of the contagious matter he was discussing, so he failed to answer the questions posed to him: How is it possible for an invisible matter to remain on the hands even after washing them with water? And how could such a small invisible matter cause the death of the patient? How is it possible that the person who did the autopsy and carried the contagion did not get infected, but instead transmitted it to the patient? It was beyond the realm of imagination for those scholars. Doctors likewise refused to accept that their hands were so unclean that people could die from contagion. With so many surgeries handled in the medical field every single day, how is it possible that autopsy would cause just one particular type of infection? Under humorism, a well-established medical system, each patient was different and thus treated accordingly. The idea of a contagion causing the same disease across all the patients would not fit. Carl Edvard Marius Levy, head of the Copenhagen maternity hospital, criticized Semmelweis's idea with these points. These were valid open questions. But in history, many discoveries have stemmed from chance accidents or unforeseen consequences, and in such cases, the investigations follow the findings. Unfortunately for Semmelweis no one thought enough of his theory to test it, and he was left alone in the battle.

Semmelweis's term at the hospital was about to expire in March 1848, roughly a year after he implemented the steps to reduce the mortality rate. Professor Johann Klein chose Carl Rudolf Braun (1822–1891) for the assistant position replacing Semmelweis. Braun was the only other applicant besides Semmelweis for the post, so it shows how much Semmelweis irked his counterparts by stressing hand washing and usage of the clean instruments. Braun was traditional in his approach to the childbed fever cases. He successfully persuaded the hospital authorities to install a new ventilation system to rid the clinic of miasmata, what he deemed the true cause of the disease. Sure enough, the mortality rate spiked soon after Braun took over the clinic, but curiously no one complained. Braun continued in that position for another five years, and by 1868 he had become the rector of the University of Vienna.

Semmelweis had applied for a private lecturer post at the university that came with limited access to the university facilities. Initially, Klein's opposition influenced the rejection of his application, and by the time it got approved in 1850, the frustrated Semmelweis had already left Vienna for the smaller town of Pest near his birthplace. He did not even inform friends about the move. One can only guess the plight of a man who knew how to save the lives of hundreds or thousands of dead mothers but couldn't because nobody would listen to him.

In Pest, Semmelweis joined a small hospital called the Szent Rokus Hospital and once again produced incredible results by enforcing the same infamous methods. Mortality rates dropped to an all time low at the hospital; in fact, only eight patients died during his six years tenure. Nevertheless, he had to face the wrath of the professor of obstetrics Ede Florian Birly (1787–1854) at the University of Pest. Birly criticized Semmelweis aggressively and firmly believed that childbed fever results from the unclean bowel. But Birly died in 1854 paving the way for Semmelweis to occupy his position. Semmelweis implemented the chlorine wash successfully, but he knew that criticism of his methods continued. He responded to Levy's criticism in an essay "The Etiology of Childbed

Fever" in 1858, followed by another essay in answer to objections raised by English physicians. And finally, Semmelweis published a major work in 1861 entitled *Die Ätiologie, der Begriff und die Prophylaxis des Kindbettfiebers* (The Etiology, Concept, and Prophylaxis of Childbed Fever). In this book, he criticized physicians for their ignorance and even pointed to the increased mortality rate at the Vienna General Hospital after his successor Braun took over. But his efforts went in vain. Even Rudolf Virchow, one of the most eminent chemists and scientists of the time, rejected Semmelweis's idea.

After all this, Semmelweis slipped into severe depression and suffered a nervous breakdown. He began drinking regularly and frequenting prostitutes. Finally, when it became embarrassing for his family and associates, they turned to Janos Balassa (1815–1868) for help. Balassa was a university professor and leading surgeon who happened to be a colleague, friend, and doctor of Semmelweis. He signed the documents referring Semmelweis to a mental asylum.

On 30 July 1865 one of Semmelweis's supporters asked him to visit one of his new institutions. Ferdinand Ritter von Hebra (1816-1880) was an Austrian physician and renowned dermatologist who founded the Vienna School of Dermatology. He had likened the results of Semmelweis to that of Edward Jenner's smallpox vaccine on the practical application of the discovery. Being an editor of a Viennese medical journal, he had published Semmelweis's chloride lime hand wash method in 1848. Now, on the pretext of taking Semmelweis to one of his institutions, Hebra instead took him to the asylum at the Lazarettgasse. Sensing the trap, Semmelweis tried to escape, but it was too late. The guards at the asylum beat him up and then put inside the asylum's dark cell. Soon after, the wounds resulted from the beating led to sepsis condition called pyemia. Ironically, Semmelweis had a section in his book 'The Etiology of Childbed Fever' that said Childbirth fever is a variety of pyemia. Two weeks later, on 13 August 1865, Semmelweis died from what the same sort of illness he had fought for his whole life. Only a handful of people attended his funeral service. Janos Diescher succeeded him at the University and implemented traditional procedures. The mortality rate jumped once again, but none of the authorities or prominent scientists took notice of it.

After many years, Semmelweis came to be known widely as the "Savior of mothers" and "*early pioneer of antiseptic procedure.*"

Semmelweis's case makes us realize that Pasteur's feat may seem like a small leap when compared with his contemporaries, but it was also an unprecedented one. Pasteur started with research on crystallography, which came to his aid while trying to decipher the puzzle of fermentation problems faced by the wine industry. And later that same work on the fermentation led him to take up the issue of spontaneous generation with confidence. This work would then steer him toward a more general theory of disease. Applying germ theory, he eventually developed vaccines for diseases like anthrax and rabies.

The work of Pasteur and those who followed him helped us look further into the true cause of disease and develop better sterilization techniques, establish safer ways to conduct surgeries, and, most importantly, save millions of lives.

Far from Dead

With his victory over Pouchet and team, Pasteur damaged the reputation of spontaneous generation theory greatly. As he put it near the end of the presentation of his experiments before the Academie members:

> *Never will the doctrine of spontaneous generation recover from the mortal blow of this simple experiment. There is no known circumstance in which it can be confirmed that microscopic beings came into the world without germs, without parents similar to themselves.*

The theory of spontaneous generation never really recovered after that. Academie's support helped in suppressing spontaneous generation proponents for the time being. Louis Pasteur was the clear winner of the decisive battle, but he could not deliver the death blow to the theory. A groundbreaking book from an obscure English naturalist would turn the debate on its head.

CHAPTER 5

THE FINAL NAIL IN THE COFFIN

On the Origin of Species by Charles Robert Darwin (1809-1882) was first published in 1859. In one relatively brief text, Darwin proposed a wide-ranging theory of evolution, and suggested natural selection as the principle mechanism with extensive evidence to support it. Evolution, as we have already seen, is strongly related to the question of the origin of life. Darwin's theory of natural selection would shape the course of scientific investigations on the subject for more than a century.

Charles Darwin

Charles Darwin

Charles Darwin was an English naturalist and geologist, born to wealthy physician Robert Darwin in Shrewsbury, England. When Darwin showed a lack of interest in medicine, his father sent him to Christ's College, Cambridge, hoping that he would become a parson. But instead, Darwin found a mentor and friend in John Stevens Henslow (1796-1861), who was an English botanist. The two became so close that Darwin was known as "the man who walks with Henslow."

When the ship HMS *Beagle* was sent out to sail South America on a two years survey, Henslow was offered the role of a naturalist aboard it. But instead, after his wife's successful protest, Henslow rejected the offer and instead referred Darwin to the ship's captain Robert Fitzroy. Darwin's father thought it would be a waste of time, but he gave approval when his brother-in-law, Josiah Wedgwood, persuaded him. So Darwin boarded the HMS *Beagle* on 27 December 1831, and would not return for five years instead of two.

On board, Captain Fitzroy gave Darwin a copy of *Principles of Geology* by well-known geologist Charles Lyell. Darwin's initial interest in the voyage was a geological survey, so the book engrossed him quickly. That book and Charles Lyell's theories had a profound impact on Darwin's thoughts.

Darwin's zeal for collecting things made him gather as many specimens as possible during the voyage. He sent back detailed notes to mentor Henslow, and as a result, when he got back from the journey on 2 October 1836, he was already a celebrity. An enthusiastic Charles Lyell met him the same month.

Evolution by Natural Selection

Charles Darwin did not conceive the idea of evolution during the voyage. In 1837, a year after the voyage, he considered the possibility of all extant species descending from one common species while investigating the vast number of specimens he had gathered. But he wasn't sure about the mechanism that could have resulted in such evolution. As Darwin recalled later in his autobiography, he had hit upon the idea of the natural selection as an explanation for the evolution while reading Malthus's "An Essay on the Principle of Population" in 1838:

> *In October 1838, that is, fifteen months after I had begun my systematic enquiry, I happened to read for amusement Malthus on Population, and being well prepared to appreciate the struggle for existence which everywhere goes on from long-continued observation of the habits of animals and plants, it at once struck me that under these circumstances favourable variations would tend to be preserved, and unfavourable ones to be destroyed. The result of this would be*

> *the formation of new species. Here, then, I had at last got a theory by which to work; but I was so anxious to avoid prejudice...*

Thomas Robert Malthus (1766–1834) was a very controversial yet influential figure in the field of political economy. He put forward the theory that the population multiplies geometrically over time, but food grows arithmetically. This will lead to the scarcity of food, as the population will eventually outgrow the food reserve. If no preventive measures are taken by the system, then natural elements like famine, disease, and wars will effectively limit population growth.

Darwin applied Malthus's line of thought to the theory of evolution. Often, there are small inheritable variations in individuals of a species. The variations that are favorable survive and are passed onto the offspring, whereas the unfavorable variations perish. And these variations make way for a new species.

It took Darwin took another four years to come up with an abstract of this theory. Darwin's efforts to publish his findings was delayed by his frequent illness after the long voyage, and also because he was trying to gather more evidence and conduct further experiments, such as the selective breeding of plants. Adding to that, his second child and the eldest daughter Anne Darwin caught scarlet fever in 1849 when she was just ten years old. Her health continued to decline as time passed, and she died in 1851.

Alfred Russel Wallace

In 1858, two long decades after Darwin had first conceived the idea of the natural selection, he received a paper from Alfred Russel Wallace (1823–1913). Darwin found himself in a tough situation as Wallace's paper was based on the same idea of natural selection. Darwin had interacted with Wallace previously as he had done with many others to discuss his findings, but he never expected to face this kind of situation. Unfortunately, Darwin was not in a position to handle the situation with a clear mind, as his youngest son was also sick with scarlet fever. So he passed Wallace's paper to Lyell, and let his friends advise him on the situation. On 28 June 1858, ten days after he received the paper from Wallace, his sick son Charles Waring Darwin died at just 18 months old.

Once all parties came to a common understanding of the situation, it was decided that both Darwin and Wallace would present a joint paper at the Linnean Society. The presentation went mostly unnoticed, but Darwin's book, published on 24 November 1859, became a huge success. Natural selection became the talk of the scientific community.

Criticism and Support

As noticed previously, Pasteur had mentioned that the theory of spontaneous generation and the theory of evolution would rule out the necessity of God. While most concurred with him on his opinion about the theory of evolution, the Church of England's response was rather mixed. As Pouchet had seen God's hand in spontaneous generation, few liberal thinkers would see God's hand in the evolution process too.

But criticism of Darwinism was not based solely on religion. Many critics complained that the nature of Darwin's theory of the struggle for existence is same as Malthusian support for capitalism. Others applied Darwinism to different fields like Herbert Spencer applied it to the society and Haeckel used it in his theory of the human race development that was racist in nature. But Darwin wanted to limit the theory strictly to the biology alone and warned not to apply it to other areas. He was even careful in mentioning that the struggle for existence between the species requires cooperation among the members of a given species.

Even Darwin's friends and staunch supporters had different opinions in one way or the other. Asa Gray, the renowned American botanist, was deeply religious and saw God as the source of all evolutionary changes. Charles Lyell, who met Darwin eagerly after the voyage of Beagle, provided strong moral support to Darwin, but he had reserved opinions towards evolution. Lyell had found the evidence for his uniformitarianism theory in Darwin's specimens, but with the evolution he saw Darwinism as a modification of Lamarckism and was extremely careful whenever he spoke about it. Within few years of the publication of *On the Origin of Species,* most naturalists believed in evolution in some form, but only very few of them trusted Darwinism. Among all his friends, Thomas Huxley was the one who supported and defended Darwinism in public on behalf of Darwin, who himself could not make many public appearances due to ill health. The passion with which Huxley would promote and defend Darwinism gained him the nickname "Darwin'sbulldog."

Thomas Huxley – Darwin's Bulldog

Thomas Henry Huxley (1825–1895) was born in Ealing, Middlesex in a middle-class family. His father George Huxley was a mathematics teacher at the Ealing School. The family ran into financial difficulties when the school was closed in 1835, which forced Huxley to drop his formal studies at an early age. Undeterred, he studied on his own and mastered theology, philosophy, geology, anatomy, and the German, Latin, and Greek languages. As Henslow influenced Darwin, a renowned British naturalist from the Isle of Man influenced Huxley. Edward Forbes (1815–1854) recognized Huxley's talent and encouraged him to go on a voyage similar to that of Darwin's HMS *Beagle*. Huxley embarked on the HMS *Rattlesnake* to northern Australia and New Guinea as an assistant surgeon between 1846 and 1850. He sent notes on the discoveries back to his mentor Forbes, who published them in England. Huxley's work mostly concentrated on the marine invertebrates. By the time he returned, he was famous at the age of only 25. The very same year he became a fellow of the Royal Society and never looked back. He became one of the finest anatomists of the time. He resigned from the Royal Navy, through which he had taken the voyage, and began an academic and scientific career. He held many positions over his lifetime in many scientific societies or associations.

Huxley had rejected Lamarck's transmutation theory of evolution for its lack of evidence. When Darwin decided to send one of the first three pre-publication copies of his work to Huxley, he had doubts about Huxley's reaction. So it was a pleasant surprise for Darwin when he received a positive response from an ecstatic Huxley. Huxley wrote later that,

> *My reflection, when I first made myself master of the central idea of the Origin' was, 'How extremely stupidnot to have thought of that!'*

Huxley never doubted the theory of evolution from then onwards, but he had some reservations about the natural selection mechanism suggested by Darwin. Nevertheless, he fully supported Darwin and warned him about the attack to follow. He asked Darwin to be prepared to face the criticism and defend evolution. In his reply to Darwin, he says,

> *"I trust you will not allow yourself to be in any way disgusted or annoyed by the considerable abuse & misrepresentation which unless I greatly mistake is in store for you— Depend upon it you have earned the lasting gratitude of all thoughtful men— And as to the curs which will bark & yelp—you must recollect that some of your friends at any rate are endowed with an amount of combativeness which (though you have often & justly rebuked it) may stand you in good stead—*
>
> *I am sharpening up my claws & beak in readiness"*

Attacks on On the Origin of Species

As Huxley predicted, there was a fierce attack on *On the*

Origin of Species. In December 1859, Huxley had published an anonymous review of Darwin's book in the *Times*. Such anonymous reviews and publications were not uncommon then. Not soon after, in April 1860, a scathing attack on natural selection ran in the *Edinburgh Review*, writtenby the finest anatomist of the day, Richard Owen (1804-1892).

Lyell had introduced Owen to Darwin during his first visit after *Beagle*'s voyage. Perhaps Owen's strong opposition to Lamarckism made Darwin refrain from discussing the theory with him. Moreover, Owen had given a talk before the BAAS (British Association for the Advancement of Science) in 1854 mentioning the impossibility of apes trans-mutating into human beings. And in 1858, Huxley had refuted Owen's claims in a talk at the Royal Institution by presenting the evidence to support the structural similarities between apes and humans.

After publishing his review attacking evolution, Owen helped Samuel Wilberforce with another lengthy anonymous review attacking Darwinism in the *Quarterly Review*. Samuel Wilberforce (1805-1873) was the lord bishop of the Oxford as well as a fellow of the Royal Society. He was an advocate of liberalism and also a great public speaker. Though a liberal himself, he could not accept evolution. Even though Wilberforce was a lord bishop and opposed the theory, the church itself took no official position on the matter.

Debate on Evolution Theory

The anonymous reviews and the debate made the theory of evolution even more sensational for the public. All these events led to a debate on evolution famously known as the Huxley-Wilberforce debate. In the BAAS annual meeting at Oxford on 28 June 1860, one paper presented by a botanist Charles Daubeny referred to Darwin's theory. Richard Owen replied by disapproving Darwin's theory by referring to his previous works describing the structural differences between the apes and humans such as the differences in the brain sizes between the gorilla and humans. Huxley was also present at the meeting, and he referred to his 1858 presentation at the Royal Society where he had refuted Owen's claims related to the transmutation from apes to humans. Finally, a debate dedicated to Darwin's theory was set to happen on 30thJune.

Huxley was not particularly looking forward to that debate. He had decided to leave Oxford the very same day. But on the 29th, a day before the debate, he happened to meet Robert Chambers and had a long chat with him. Robert Chambers (1802-1871) was a geologist and publisher who supported Darwin's theory. Even before Darwin's work had been published, Chambers was successful in creating a sensation about the topic. He had published *Vestiges of the Natural History of Creation* in 1844, suggesting a cosmic evolution theory and linked it with the Lamarck's transmutation theory. As he had rightly predicted, the book created an uproar and garnered harsh criticism from the intellectuals of the time who had rejected Lamarck's transmutation theory. Thankfully for Chambers, he had chosen to publish anonymously. Nevertheless, it created a stir and became a bestseller of the time. But nobody could identify Robert Chambers as the author until after his death.

After desperate persuasion from Chambers, Huxley decided to stay and participate in the debate, unknown at the moment that this event would mark a permanent place in the history of science. That day, more than a thousand people gathered to witness the arguments. Darwin's old mentor, Henslow, chaired the debate. Though Henslow had been Darwin's mentor previously, he disapproved of *On the Origin of Species* mostly for the theological reasons. Darwin, being ill, couldn't attend the debate. After many presentations, it was Wilberforce's turn to present his argument. The night before, Owen had coached him and provided the scientific points opposing Darwin's theory. With that help, and being a great orator, Wilberforce effectively presented his stand against evolution. Referring to Huxley's comments related to the structural links between the gorillas and humans, Wilberforce provoked Huxley by asking if he had a preference for the descent from ape being on his mother's side or his father's side. And in his famous reply, during his turn to speak, Huxley responded that he would not be ashamed to accept that the ape was his ancestor, but he is ashamed to be linked with a man who uses his excellent oratory skills to try and suppress the truth. Though Huxley was still very young and lacked maturity in oratory skills, his remark had a great impact on the public. After Huxley and Wilberforce, even Fitzroy, who was the captain of the ship that significantly shaped Darwin's work, spoke. Fitzroy strongly opposed Darwinism. And adding to it, he felt guilty for contributing to that theory by taking Darwin with him on the ship. Lifting a Bible in his hands, he asked the public to believe in God and not man. After Fitzroy, Darwin's other close friend, botanist Joseph Dalton Hooker, put forward scientific points and successfully countered Wilberforce's objections.

The full debate is not documented officially, so most of the details come from the secondary sources, like what the witnesses and participants have written or claimed afterward. Some witnesses claimed that at the end of the debate "Soapy Sam," as Samuel Wilberforce was called popularly, had clearly lost. But the letters and communications between the involved parties from both the circles indicate that both the groups thought their side had the upper hand at the end. Nevertheless, this debate made Huxley a familiar name in public.

It is also interesting to note that Wilberforce maintained a good relationship with both Huxley as well as Darwin after the debate. But Owen never reconciled with Huxley and Darwin. Even though the history remembers Wilberforce mostly for this debate alone, he was a liberal and reformist in his field. His father William Wilberforce was a champion of abolitionism who had successfully led campaigns to ban the slavery completely in 1833. Samuel Wilberforce's opposition to Darwin's theory was mainly triggered by the input of the scientific minds like Owen and from the fear of its social implications. But he held no personal grudge against Darwin.

Richard Owen, to the contrary, took the debates personally. Richard Owen was a renowned anatomist. He had coined the term "dinosaur" in 1842 and even added the first specimen of Archaeopteryx to the natural history museum in 1863. But Huxley too had become a fine comparative anatomist of the day. Huxley had concluded that birds evolved from the dinosaurs after studying archaeopteryx. But Owen did not receive Huxley's anatomical studies and evidence in the right spirit. Owen remains a controversial figure due to his involvement in many such incidents. Before Darwinism, Darwin and Owen had a strictly professional relationship. After Darwinism's success and popularity, Owen became one of Darwin's harshest critics . As Huxley reminded Wilberforce in the debate, in the questions about science, the outcome of scientific investigations will always be the real authority. If one sticks to his personal views even when the investigation suggests otherwise, then he is doomed to lose the battle.

Genesis

While Huxley is often called "Darwin's bulldog," he did not support Darwin's theory blindly. Even though he was satisfied with the evidence in support of the evolution, he could not agree entirely with Darwin on the mechanism of evolution: natural selection. Huxley was an empiricist who believed in the empirical evidence obtained through the scientific inquiry, and he saw little evidence to support natural selection. He expressed this to Darwin, and Darwin responded that it was impossible to witness the transformation of one species into another, but that natural selection provides an answer to most of the questions. But Huxley remained agnostic about it. He coined the term "agnosticism" to indicate a position that does not pretend to know, or not to know, something for certain that is not yet determined or determinable. Not just natural selection, he applied the same term to clarify his position about God as well. He did not claim to be an atheist who knew for certain that there was no God. At the same time, he did not claim that God exists because he did not find enough evidence to prove it.

For the same reason, Huxley always stressed the importance of keeping science and religion separate. And so, Huxley did not approve of the ending note added by Darwin in *On the Origin of Species* attributing the origin of first life to some "grandeur" and using terms from the Book of Genesis -

> *"...There is grandeur in this view of life, with its several powers, having been originally breathed into a few forms or into one; and that, whilst this planet has gone cycling on according to the fixed law of gravity, from so simple a beginning endless forms most beautiful and most wonderful have been, and are being, evolved."* - Darwin, Charles (On the Origin of Species By Means of Natural Selection, or, the Preservation of Favored Races in the Struggle for Life)

As Huxley, and later others pointed, this gave the suggestion that God (a mysterious "grandeur") breathed the first life on Earth, and then evolution followed. Huxley told Darwin that he should have given up the discussion on the origin of first life altogether or accepted the necessity of spontaneous generation. Darwin insisted that he did not mean it the way others had taken it. As Huxley said in the letter to Charles Kingsley (1819-1875), an author, professor, priest and a friend of Darwin as well as Huxley, on 22 May 1863,

> *The x of science and the x of genesis are two different x's, and for any sake don't let us confuse them together.*

Darwin had an answer to the appearance of all the life forms that came after the first life. But he had no answer for the first life on Earth. He did not believe the theory of spontaneous generation that suggested putrefaction leads to life, but when he had formulated the theory of evolution in his mind in 1837, he was ready to agree to the possibility that life might have originated from some simple inorganic matter. He wrote in his notes that,

> *The intimate relation of Life with laws of chemical combination, & the universality of latter render spontaneous generation not improbable*

But by the time he published *On the Origin of Species*, his opinion on the origin of life was changed. He now had even less favor to the idea of the spontaneous generation, possibly due to the influence of growing debates like the one between Pasteur and Pouchet. So Darwin refrained from discussing spontaneous generation in public. But his reluctance to discuss the origin of life question, as well as attributing the cause of it to some "grandeur" in the book, brought him harsh criticism.

For many, the lack of an answer to the origin of life meant a weaker theory of evolution. And many materialists complained that his reference to some "grandeur" was a move towards the spiritualism. One such criticism came from Heinrich Georg Bronn (1800-1862). Bronn was a reputed German paleontologist and geologist even before he started working on the three-volume *Handbuch einer Geschichte der Natur* (Handbook of a history of nature) in the 1840s. He connected multiple branches of science like zoology, geology, cosmology, and paleontology and tried to establish how the Earth came into being and how organic life changed over the ages. He sought to identify and establish the common laws governing those changes. While discussing how the different species came into existence, he disagreed with the theory that one species could transform into a different species. Instead, he attributed it to some creative force that would produce new species at specific interval suitable for the environment.

So when Darwin came up with his theory of evolution, Bronn did not agree with it. Still, he became the first person to translate Darwin's book into a different language (German) and promoted the theory. While he promoted Darwinism, he made sure to include his objections by writing an essay and even adding a commentary at the end of the book. One of his main criticisms was aimed at Darwin's failure to answer the question of the origin of life. Darwin had left a door open for the existence of mysterious force by not answering how life began. This, Bronn said, makes Darwin's theory incomplete.

Darwin strongly disagreed with this criticism. As per him, the origin of life, though important for evolutionary theory, was an entirely different subject. One need not have an answer to the life's origin to prove that evolution occurred. He wrote to Lyell on 16 February 1860 expressing the same about Bronn's criticism,

> *He...seems to think that till it can be shown how life arises, it is no good showing how the forms of life arise. This seems to me about as logical (comparing very great*

things with little) as to say it was no use in Newton showing laws of attraction of gravity & consequent movements of the Planets, because he could not show what the attraction of Gravity is.

Richard Owen, relentless, wrote another anonymous review published on 28 March 1863 by the *Athenaeum*, a highly reputed magazine of the time in England. This time Owen attacked Darwinism while reviewing the physician William B. Carpenter's "Introduction to the study of Foraminifera." Carpenter had linked the existing and extinct forams(Foraminifera) by identifying strong similarities between the primitive and extinct forams and the existing ones. Though Carpenter supported evolution, he saw this example as an objection to Darwinism because the forams had not seen a significant change in a very long time . If Darwin's theory were true, surely they would see far more changes in this species. Owen, reviewing the study, criticized both evolution and Darwinism for being weak, as clearly forams did not go through the evolution process. While doing so, he also supported heterogenesis, mentioning that the foraminifera and some other organisms are generated through heterogenesis process "*during each geological period in slimy ooze.*" Also, Owen attacked Darwin by remarking that the use of "Pentateuchal terms"—by which he was referring to the usage of the term "grandeur," —showed that the evolution theory is weak.

Hooker forwarded a copy of this review to Darwin. Darwin replied to him on 29th March,

I have long regretted that I truckled to public opinion,

> *and used the Pentateuchal term of creation, by which I really meant "appeared" by some wholly unknown process. It is mere rubbish, thinking at present of the origin of life; one might as well think of the origin of matter.*

This time Darwin responded, and *Athenaeum* published his reply on 18th April. Darwin asked if we were to consider heterogenesis as true, as the reviewer (Owen) had claimed in his critique, then could we consider that the first life on Earth has originated spontaneously from purely inorganic matter and natural forces? Darwin says, "*At present it is to us a result absolutely inconceivable.*" Finally, he clarifies the usage of "Pentateuchal terms,"

> *...in a purely scientific work I ought perhaps not to have used such terms; but they well serve to confess that our ignorance is as profound on the origin of life as on the origin of force or matter.*

Then Darwin moved to answer the Carpenter's objection that the foraminifera did not change as much as one would expect, the details of which will be discussed in a more relevant section.

Interestingly, though Pasteur had won the first round of the hot debate on spontaneous generation less than a year ago, neither Owen nor Darwin mention his name. But between 1837 and 1863, Darwin's view of spontaneous generation had changed. Perhaps the cases where his theory had been attacked for using the spontaneous generation as an answer for the origin of first life or the debates between Pasteur and

Pouchet forced Darwin to change his mind. In his letter to Professor Julius Victor Carus (1823-1903), German comparative anatomist who had become Darwin's friend and translator of all works starting with the 3rd edition of *Origin*, on 21 November 1866, Darwin asserts his position,

> *As for myself I cannot believe in spontaneous generation, and though I expect that at some future time the principle of life will be rendered intelligible, at present it seems to me beyond the confines of science.*

Nevertheless, Darwin was always intrigued with the question of the origin of life. With time his opinions seem to change again. As many events unfolded later, he was willing to consider the possibility of some form of spontaneousgeneration explaining the origin of life. It is not clear whether the changes in his religious views had anything to do with this, but one cannot rule out that possibility.

The History of Creation

One great admirer of Charles Darwin was disappointed with him for not dealing with the question of the origin of life. Ernst Heinrich Philipp August Haeckel (1834–1919), a German who was into many fields like biology, philosophy, and art, was quick in accepting and promoting Darwin's theory, but he expressed dissatisfaction as well:

> *The chief defect of the Darwinian theory is that it throws*

no light on the origin of the primitive organism—probably a simple cell—from which all the others have descended. When Darwin assumes a special creative act for this first species, he is not consistent, and, I think, not quite sincere. However...

Haeckel was a man of many talents. He has contributed significantly to different fields, but he also remains a controversial figure because of ideas about race and social Darwinism. Haeckel identified ten distinct human races and ordered them based on which race was more civilized. To him, white Caucasians were the most civilized, and Negroswere closer to apes than any other race. He based this view on the observations made from his travels and his close interactions with native peoples. He also supported polygenism, which suggests that the different human races have different origins. In each such origin group, the primitive organisms underwent an evolution process in a Lamarckian way, leading to different human races. This had an obvious implication that not all races are equal. Haeckel became a leading figure in pseudoscientific racism by trying to apply biology concepts to the society and politics. With the publication of *Riddle of the Universe* in 1899, he also became a leading figure in social Darwinism. It is important to note that even though the ideologies of social Darwinism and eugenics, which apply the concept of Darwinism to the society, carry Darwin's name heavily, Darwin himself never supported them. Instead, he warned strongly against applying his theory to the society.

Nevertheless, Haeckel defended the natural selection theory with passion. Still, as his notes indicate, he criticized Darwin for trying to evade the question in his otherwise an exemplary work. Darwin was too cautious to jump to any conclusions without enough evidence, but Haeckel did not have that problem. He would frequently formulate a stronghypothesis based on the available scientific facts and fill the gaps with suppositions. His brilliant mind produced many hypotheses that proved true at a later point in time. And Haeckel had a theory about the origin of life as well.

Haeckel did not bother himself with spontaneous generation debates like the one between Pasteur and Pouchet. He also ignored heterogenesis because it does not help answer the question of first life. The very first life must have come from the inorganic matter alone, and Haeckel dealt with that problem. He discussed the subject of the origin of life in *Natürliche Schöpfungsgeschichte* published in 1868. The same was translated into English in 1876 as *The History of Creation* in two volumes.

Oken and Lamarck greatly influenced Haeckel. He linked Oken's theory of the origin of life with Lamarck's evolution theory and used the latest scientific achievements to develop them further. Oken had come up with the concept of Ur-Schleim about sixty years back. Ur-Schleim was nothing but a very basic mucus mass, and the agglomeration of the Ur-Schleim results in all the complex organic bodies. Later Schwann and Schleiden developed their cell theory, which states that all organic life is made up of cells. So, per Haeckel, the cell theory supported Oken's theory. After the origin of first life, we needed to explain how all the different life forms came about. Lamarck answered this by explaining how all the life forms have evolved, but we still needed to understand the mechanism of evolution. And Darwin had just done that with the theory of natural selection. Darwin had even come up with the evidence for evolution that Lamarck lacked.

Lamarck explained the origin of life with the steady state theory where the spontaneous generation from the inorganic matter is followed by the evolution process to give rise to different life forms constantly and continuously. But, Haeckel complains about Darwin,

> *Darwin passes over and avoids this subject, as he expressly remarks that he has "nothing to do with the origin of the soul, nor with that of life itself." At the conclusion of his work he expresses himself more distinctly in the following words:—"I imagine that probably all organic beings which ever lived on this earth descended from some primitive form, which was first called into life by the Creator."*

By attributing the origin of life to a creator, Haeckel says, we "*pass from the domain of true science to the completely distinct domain of poetical faith,*" and before taking such a decision, we are "*in duty bound to endeavor to examine it in the light of a mechanical hypothesis.*"

In *The History of Creation*, Haeckel first tries to establish the connection between organic and inorganic matter. When we compare the chemical elements in organic and inorganic matter, we find no single "special" or "different" element in the organisms when compared to inorganic matter. So when the nature of elements remains the same, where is the difference? It is in how the chemical bonds unite these elements. The carbon combining with other elements like hydrogen, oxygen, and nitrogen forms the compounds that are "albuminous" in nature. The living organisms have a "*firm-fluid or swollen state of aggregation*" (viscid) that lies somewhere between the solid and liquid states, whereas inorganic matter has only three states– solid, liquid and gaseous.

Haeckel says, the simplest organism is nothing but "*semi-fluid albuminous lump*" and is called the "*Monera.*" At a later stage, the cell nucleus or kernel forms in the interior of Monera separating from the surrounding albuminous matter called the "protoplasma," thus forming the cell. And the protoplasma undergoes gradual adaptation in different cells in many ways, resulting in the different types of cells. All organisms are nothing but the aggregation of such different forms of cells. Thus, stressing the vast similarities between the properties of organic and inorganic matter, Haeckel declares that,

> *...we can no longer draw a fundamental distinction between organisms and anorgana, a distinction of which, formerly, naturalists were generally convinced.*

Haeckel distinguishes spontaneous generation into two types: autogeny and plasmogeny. Autogeny is where the organism generates out of the inorganic matter; there is no organic matter involved in this process. Plasmogeny is where the organism generates from matter that includes some organic matter. In essence, Haeckel's plasmogeny was nothing but what others called heterogenesis, and it had no significance when it comes to the origin of life. Haeckel agreed there was no definitive evidence available for both autogeny and plasmogeny yet. Though many experiments had been done by many scientists, it was still not conclusive. But, Haeckel says, when some experimenters claim that the negative results obtained in their experiments prove that spontaneous generation does not occur, it is not true. Instead, their failed experiments prove only that the conditions used in their experiments failed to produce spontaneous generation. The conditions on the primitive Earth differed greatly from the current times, and one cannot prove that those conditions were not favorable for the spontaneous generation. So even if we have no conclusive evidence at present, spontaneousgeneration must still remain a supposition to us.

Haeckel supported the origin of life through autogeny or spontaneous generation, but he did not accept the role of any supernatural force in it. Renowned chemists and other scientific minds who supported vitalism said that one couldn't produce the organic matter from inorganic matter because organisms contain a vital force. Haeckel points that, starting with Wohler in 1828, many others have proved it wrong by synthesizing organic compounds from inorganic material. This rules out the need for any mystic or supernatural force for spontaneous generation to occur.

After strongly establishing the possibility of the spontaneous generation, Haeckel then dives deeper into the details of the origin of life. As per the cell theory, all organisms are made up of cells, but cells are complex in structure with a nucleus and the cell membrane so they must not have come from spontaneous generation. Only the simplest albuminous homogenous bodies containing protoplasma , or Monera, could have come from such a process. Except for the fact that these Monera exhibit two fundamental functions, "'nutrition" and "propagation," of any living organism, they act almost like inorganic matter. And there are seven genera of Monera like the protamoebae from kingdom Protista. It was Haeckel who identified and suggested the kingdom Protista separating it from plants and animals.

Haeckel refers to the basic units of life as the form-units or plastids, and he classified them into four categories –

1. *The simple cytodsor Gymnocytoda*: These simple cytods are nothing but the simple Monera consisting only of the protoplasma. These will not contain cell membrane or nucleus. And these are the only basic form-units that are generated by spontaneous generation.
2. *The encased cytodsor Lepocytoda*: When the simple cytods undergo the process of adaptation through natural selection, a chemical agglomeration or some condensation causes a membrane to form around the protoplasma. This membrane will protect the cytod's interior from any harmful external factors. These new cytod forms are called the enclosed cytods.

- *The simple cells or Gymnocyta*: Sometimes a condensation process happens at the innermost part of the protoplasmaof a simple cytod, thus forming a kernel or nucleus at the center. These are simple cells with no membrane but having a nucleus at the center.

1. *The encased cells or Lepocyta*: When condensation at

> the center of simple cytod forming the nucleushappens along with the condensation or chemical accumulation at the border forming the cell membrane, then we get enclosed cells. These have both nucleus as well as a membrane.

These four basic form-units would follow the process of adaptation as explained by Darwin, leading to every other life form.

But if the Monera went through the evolution process to form other species, then why do we still see them? There could be two reasons for this: First, Monera follow the reproduction process like any other organisms, and those that have not transformed into another species since the ancient times would continue to exist through propagation without losing their original structure. Second, Monera are being generated all the time spontaneously. And the possibility of latter is more likely than the former.

Haeckel agrees that his theory is purely hypothetical with no evidence to support it, but he stresses that it is "*indispensable for the consistent completion of the non-miraculous history of creation.*" Leaving the reader an option to choose between his hypothesis and the supernatural creator, Haeckel mentions,

> *...the idea that the Creator should have in this one point arbitrarily interfered with the regular process of development of matter, which in all other cases proceeds entirely without his interposition, seems to be just as unsatisfactory to a believing mind as to a scientific intellect. If, on the other hand, we*

assume the hypothesis of spontaneous generation for the origin of the first organisms, which in consequence of reasons mentioned above, and especially in consequence of the discovery of the Monera, has lost its former difficulty, then we arrive at the establishment of an uninterrupted natural connection between the development of the earth and the organisms produced on it...

In his second volume, Haeckel discusses the monophyletic versus polyphyletic origins hypothesis. While monophyletic hypothesis suggests that every organism came from a single common ancestor, the polyphyletic hypothesis suggests that different groups of organisms came from different common ancestors generated spontaneously at the same time. Even though Haeckel mentions that it is impossible to prove which hypothesis is correct, he concludes by favoring the polyphyletic hypothesis.

While discussing the different living Monera, Haeckel reports the discovery of "Bathybius Haeckelii" by Huxley, who had named it in honor of Haeckel. Huxley had found it in 1868 in samples from deep ocean sediments, and he was surprised to see how similar it was to Haeckel's description of the Monera. Haeckel was excited to get the report of this discovery from Huxley and quickly added it to his list of Monera. But later Huxley figured that what he'd found was just a chemical substance, and immediately accepted the mistake.

This incident is a mirror to the very nature of Huxley. If the empirical evidence was against his theory, he was open to accept it and reject the theory, no matter the consequences. He expected the same from all his colleagues and students who worked with him. He would distance himself from the people who stuck defending their opinions even when the evidence is contrary to it.

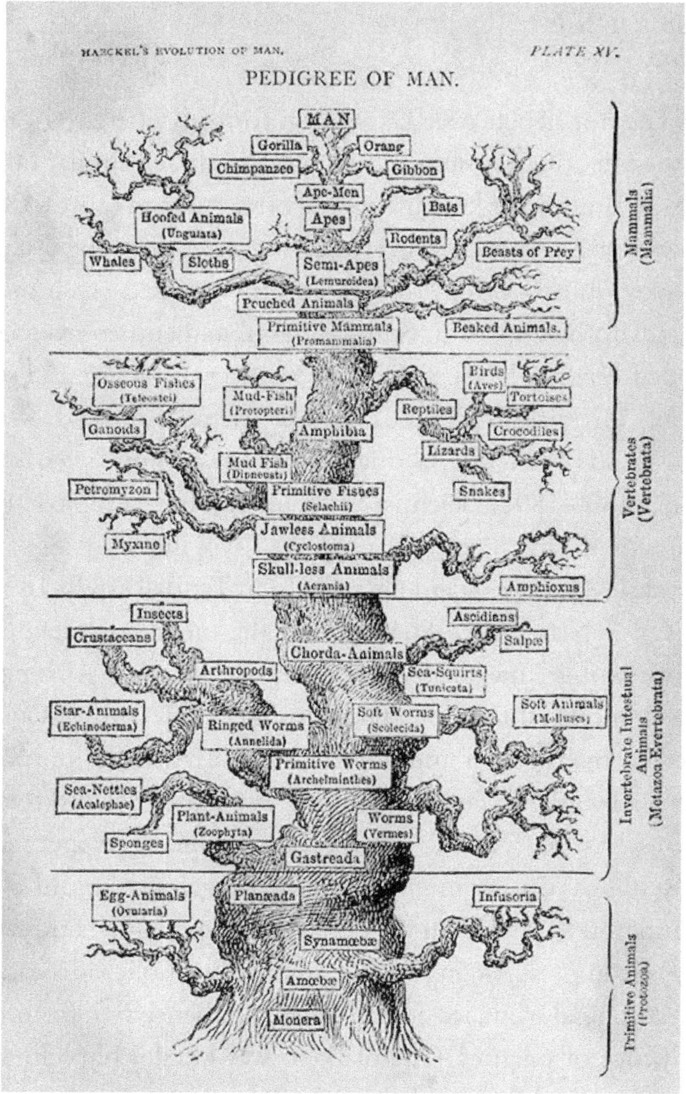

Haeckel's 'Pedigree of Man' from The Evolution of Man, vol ii

The X Club

The hot debate over Darwinism formed different groups in the scientific community. Those who shared similar opinions became closer knit and would often meet to express their views. On such occasion, Huxley suggested forming a club. Hooker quickly agreed to this, and they invited seven other friends to be part of it. So the final nine-member group included Huxley, Hooker, Herbert Spencer, zoologist and surgeon George Busk (1807-1886), chemist Sir Edward Frankland (1825-1899), mathematician Thomas Archer Hirst (1830-1892), scientist Sir John Lubbock (1834-1913), mathematician and physicist William Spottiswoode (1825-1883), and physicist John Tyndall (1820-1893).

On 3 November 1864 eight of the members, excluding Spottiswoode, met at the St. George's Hotel on Albemarle Street in central London. The purpose of this club was just a social come together and served no other particular purpose. After a lengthy discussion over the name, they decided to call themselves the "X Club." After that, the members met regularly, almost once a month. As time progressed, so did their reputation in their respective fields. Some members played a key role in popularizing science among the public, advocating for empirical evidence and for keeping science free from the influence of religion. One of them was Huxley himself, and the other was John Tyndall.

When John Tyndall, being a physicist, supported the theory of evolution it was a happy surprise for Darwin and his fellow naturalist friends. When Hooker expressed the same to

Darwin, he replied with similar expression on 6 August 1871 –

It is a gain that so wonderful a man, though no naturalist, should become a convert to evolution...

John Tyndall was born to a poor family in Leighlinbridge, County Carlow, Ireland. Between 1842 and 1847 he worked for the railway companies in planning new rail constructions. The concept of the rail construction grew since the first decade of the 19th century, and in the 1840s it was fully thriving. The railroad industry grew so much that almost all the towns in England had railway network by the end of that decade. And in such good times, Tyndall was naturally paid well. He quit the job in 1847 after saving some money and joined Queenwood College as a teacher. There he met Edward Frankland, who would later become a fellow X Club member. Both friends pursued their interest in science and joined the University of Marburg in Germany. Tyndall studied under Robert Wilhelm Eberhard Bunsen (1811-1899), a German chemist known for developing the Bunsen burner among many other things, and Karl Hermann Knoblauch (1820–1895), a physicist known for his contribution towards the study of infrared radiation. After learning from such great minds,Tyndall was now ready to make his own reputation.

Tyndall's work on the diamagnetism in collaboration with his mentor Knoblauch brought him recognition in the field. By 1852, just a year after his return to England, he had become a fellow of the Royal Society. And within next year he joined the Royal Institute in England as a professor of Physics. Already a leading authority in the same field, Michael Faraday (1791-1867) at the Royal Institute was impressed with Tyndall's work. They became good friends and colleagues, and Tyndall even published "Faraday as a Discoverer" in 1868 in Faraday's memory.

Tyndall became a leading figure in popularizing science . He gave hundreds of open lectures for the public and ensured that the scientific concepts and theories reached the laymen effectively. As part of the same effort, Tyndall published several bestselling books and wrote in journals. Inevitably, he met Huxley in that endeavor and both became good friends. Tyndall, like Huxley, wanted to keep science free from the influence of religion.

Tyndall was involved in the studies of the radiant heat and its effects. He proved the greenhouse effect for the first time. He noticed the suspended particles in the air matter while working on the effects of radiant heat and its relation tothe gaseous matter. He decided to filter those particles as he needed particle-free air in his work. In his efforts towards achieving the same, he deduced that a strong beam of light passed against the dark environment is effective in identifying those suspended particles in the air and water, just like how one would notice the particles in a dark room through a beam of light entering it. Further, he was surprised when those particles passed freely through "*alkalis, acids, alcohols, and ethers.*"

Unlike many other physicists, Tyndall's interest was not just limited to his academic or research work. The debates on the spontaneous generation and evolution interested him greatly. So when he noticed that the air had a lot of floating particles that could pass through many solutions, he was able to quickly relate it with the previous experiments done by various people on the subject of spontaneous generation. Schulze, Schwann, Schroder, and Dusch had done similar experiments, as had Pasteur. Tyndall decided to repeat those experiments and confirm the presence of particles in the air during putrefaction. And he concluded that whenever there was putrefaction, there were floating particles in the air, and when there was no putrefaction, it was completely free of any such particles. Tyndall put his results before the members of the Royal Institute in 1870, and once confident enough he went public by publishing the findings in the *Times*.

As Pasteur faced the criticism by Pouchet, Tyndall was also criticized by someone he did not expect—Henry Charlton Bastian (1837–1915), a physician who rose to fame at a very young age. As Tyndall himself recalled later in his book *Floating Matter of the Air* about Bastian's criticism,

> *This letter* (to the Times), *I believe, gave occasion for the first public utterance of Dr. Bastian in relation tothis subject. He did me the honor to inform me, as others had informed Pasteur, that the subject "pertains to the biologist and physician." He expressed "amazement" at my reasoning and warned me that before what I had done could be undone "much irreparable mischief might be occasioned." ... As regards the public who take an interest in such things, and apparently also as regards a large portion of the medical profession, our clever countryman succeeded in restoring the sub-*

ject to a state of uncertainty similar to that which followed the publication of Pouchet' volume in 1859.

Bastian had an excellent academic record. He graduated from the University of London in 1861. By 1866 he was certified as an MD from the same university, and within next two years he became a fellow of the Royal Society. After his studies, he worked as a lecturer in pathology at the St. Mary's Hospital Medical School, and in 1868 he joined the National Hospital as one of its first neurologists, where he would serve until the end of his career.

When he was still aiming for MD in 1865, Bastian wrote a thesis on nematoid worms in which he had identified more than 100 new species. This quickly garnered attention, including from Royal Society members. Huxley and Busk reviewed Bastian's thesis to decide if it was good enough to be added to the *Philosophical Transactions*, a journal published by the Royal Society. Both were extremely impressed with the quality of that thesis. While they approved the paper for inclusion, they also suggested adding Bastian as a member of the Royal Society. As a result, Bastian became Fellow of Royal Society in 1868.

Later Bastian worked on the speech disorders, and his extensive works on aphasia would mark his name permanently in the history. He became a fervent proponent of evolution by natural selection after reading Darwin's book. The passion with which he defended the evolution and the recognition he had received at an early stage of the career made him one of the leading figures of evolutionary theory in the younger generation. Not surprisingly, Darwin and the X Club members, including Tyndall, were very happy to have him on their side—that is, until Bastian began to devote all his energy to the question of the origin of life and the role of spontaneous generation.

After Tyndall's presentation at the Royal Society on the air particles during the putrefaction process, X Club members discussed it at their next meeting. After the meeting, nearly everyone favored Pasteur's germ theory and Tyndall's experiments. Endorsing germ theory did not mean they rejected the possibility of the life arising from chemicals in the beginning. Being followers of Darwinism, they knew that such an event triggering the origin of first life on Earth was necessary, but it does not mean that the life is continuously and spontaneously generated as Pouchet had believed.

Huxley retained his "agnostic" position regarding the origin of life, but his recent works such as the discovery of "Bathybius Haeckelii" and his public statements made people think that he favored spontaneous generation. Huxley got in touch with Bastian after reviewing his paper for *Philosophical Transactions* and also petitioning for his membership to the Royal Society. So when Bastian decided to work on this subject, he approached Huxley. Huxley encouraged him to proceed with it and offered his help and advice whenever needed.

When Tyndall presented his experiments before the Royal Society, Bastian was already working on the subject. He had begun his work not with an intention to test if spontaneous generation is true, as he was convinced of its existence, he was aiming to prove it to the world. His thought process was in line with Haeckel's. Being an ardent believer in evolution, he knew that life must have begun spontaneously the first time. It was imperative for the theory to work. Haeckel had come up with the theoretic approach that favored the spontaneous generation of simplest protoplasma or "Monera." Similarly, Bastian formed an opinion that first life must have originated spontaneously, and it must be happening all the time as we can still find those small organisms. These organisms, being generated spontaneously, cannot result from evolution but instead must be the source material for the evolution process.

Sphagnum Moss Incident

While Bastian worked furiously to prove spontaneous generation, it must have come as a surprise X Club member Tyndall came out in support of germ theory and against spontaneous generation. Considering that Tyndall was not just any other scientist but a celebrated popular figure, Bastian rightly assumed that Tyndall's work would cause significantdamage to the theory of spontaneous generation in public opinion. So he sent a sharp rebuke to the *Times*.

The very same month, April 1870, Bastian told Huxley that his experiments had provided clear evidence supporting spontaneous generation. Huxley already knew about Tyndall's experiments and was impressed with their quality. And after looking at Bastian's experiments, he could not say the same. As Huxley would recall it later, the very first specimen Bastian presented had a leaf of Bog Moss (Sphagnum), and he had a hard time convincing Bastian about it:

> *...when the first of these wonderful specimens was put un-*

der my microscope, I told him at once that it was nothing but a fragment of the leaf of the common Bog Moss (Sphagnum); he will recollect that I had to fetch Schact's book "Die Pflanzeuzelle," and show him a figure which fitted very well what we had under the microscope, before I could get him to listen to my suggestion; and that only actual comparison with Sphagnum, after he had left my house, forced him to admit the astounding blunder which he had made.

Though Huxley concluded that the quality of Bastian's experiments could not be trusted, he did not wish to hurt the feelings of such a brilliant young man. Huxley took extreme care in explaining his observations and points. To his credit, Bastian took the advice sportingly, although he didn't necessarily take it to heart.

After going public criticizing Tyndall's work, Bastian wanted to publish experimental results as soon as possible. But he could not get a slot to present to the Royal Society, so he decided to go public once again. Huxley's opinion on Bastian's experiments at the time was completely shaken. Huxley was surprised by Bastian's tone of confidence even while failing to identify the clear errors in his work. On 30 April 1870, Huxley mentioned Bastian in his letter to Felix Anton Dohrn (1840-1909), a German zoologist and Darwinist, -

"... Then there is our Bastian who is making living things by the following combination:—

Prescription: Ammoniae Carbonatis
Sodae Phosphatis
Aquae destillatae

quantum sufficit
Caloris 150 degrees Centigrade
Vacui perfectissimi
Patientiae.

Transubstantiation will be nothing to this if it turns out to be true, and you may go and tell your neighbourJanuarius to shut up his shop as the heretics mean to outbid him."

The very next day, on 1 May 1870, Huxley warned Bastian not to go public until he rectified all the errors in the experiments and confirmed their reliability. But Huxley's concern for Bastian's feelings only made the situation worse. Bastian thanked Huxley for the advice and excluded the experiments that Huxley had pointed out, but failed to understand the real caution behind Huxley's words.

Huxley was not the only voice in Bastian's ear, though. Bastian was encouraged by English astronomer Sir Joseph Norman Lockyer (1836-1920) to publish his works in a new scientific journal called *Nature*. Lockyer had contacted the respected publisher Alexander Macmillan and convinced him to print the new journal, and both Lockyer and Macmillan were impressed by Bastian's work. As a result, Bastian's paper appeared in *Nature* in three parts starting on 30 June 1870. Lockyer and Macmillan would continue to support Bastian.

Huxley was disappointed with Bastian because of the publication. Presenting the poorly done experiments as clear evidence for spontaneous generation was one thing, but now Bastian had published his experiments in a hurry without taking Huxley's advice seriously. After this incident, Huxley did not validate Bastian's experiments again. He did not go public refuting Bastian's claims either; instead, he repeated the related experiments by incorporating Tyndall's method. He found no evidence of spontaneous generation, which convinced him that Bastian's experiments must be erroneous.

Darwin and the X Club members discussed Bastian's paper internally. On 6th July Hooker wrote to Darwin:

> *Bastian's paper in Nature is full of curious matter but eminently unsatisfactory in treatment I think, & poorly written.*

But by then Huxley had informed them about Sphagnum leaf incident, so Darwin replied Hooker next day mentioning the same:

> *When you see Huxley ask him about Bastian & a fragment of Sphagnum moss, which he mistook for a spont. gen. organism!*

On 12th July Darwin sent another letter to Hooker where he continued the discussion:

> *Spontaneous generations seems almost as great a puzzle as preordination; I cannot persuade myself that such a multiplicity of organisms can have been produced, like crystals, in Bastian's solutions of the same kind. I am astonished that as yet I have met with no allusion to Wyman's positive statement that if the solutions are boiled for 5 hours, no organisms appear; yet, if my memory serves me, the solutions when opened to air, immediately became stocked. Against all ev-*

idence I cannot avoid suspecting that organic particles (my gemmules from the separate cells of the lower creatures!) will keep alive & afterwards multiply under proper conditions. What an interesting problem it is.

Wyman referred here was Jeffries Wyman (1814-1874), an American anatomist and Hersey professor of anatomy at the Harvard College. He had experimented on heating of the organic matter solutions and published a book called *Observations and Experiments on Living Organisms in Heated Water* in 1867. Below are some conclusions drawn in his book:

"2d. Solutions of organic matter boiled for twenty-five minutes, and exposed only to air which had passed through iron tubes heated to redness, became the seat of infusorial life. Exps. I-V.

3d. Similar solutions contained in flasks hermetically sealed, and then immersed in boiling water for periods varying from a few minutes to four hours, also became the seat of infusorial life. The infusoria were chiefly Vibrios, Bacteriums and Monads. Exps. VI-XV.

4th. No ciliated infusoria, unless Monads are such, appeared in the experiments referred to in the above conclusions.

5th. No infusoria of any kind appeared if the boiling was prolonged beyond a period of five hours."

Bastian had noted that infusoria would appear even in the solutions hermetically sealed and subjected to prolonged heat.

Darwin suspected that the organisms most likely remained alive even after the heating, and later when subjected to the cooling conditions they multiplied. Even though it was just a suspicion, this was an important idea that would resurface later.

Abiogenesis

Huxley became convinced that Bastian must have erred in his experiments. But he did not try to convince Bastian because he doubted if Bastian would ever listen. Instead, he thought of clarifying his stand on the matter based on the available empirical evidence. He got that opportunity a few months later at the 40th annual meeting of BAAS in Liverpool on 14 September 1870.

In a highly impactful presidential speech to the conference, Huxley laboriously detailed the history of spontaneous generation research, including the works of Harvey, Redi, Schulze, Schwann, La Tour, Helmholtz, Schroeder, and Dusch. Finally, he lauded Pasteur for his groundbreaking work and Tyndall for his recent experiments validating the work of those who came before. Huxley also introduced some new terms and attempted to clarify meaning for some terms that were used by different people in different ways. He coined the term "abiogenesis" to indicate living matter coming from non-living matter. The opposing term "biogenesis" indicated the living matter generating from existing living matter. Huxley then defined two forms of biogenesis: "homogenesis"—parents giving birth to an offspring of the same form—and "heterogenesis"—parents generating offspring in an entirely different form. Huxley's usage of these terms became standard thereafter, though Bastian still preferred to use the term "archebiosis" instead of "abiogenesis."

Mentor to Rival

Even though Huxley did not mention his name anywhere in the speech, Bastian was hurt by Huxley's stand on the subject. Bastian thought that, like Tyndall's paper, Huxley's speech would be damaging for the theory of spontaneous generation in public. Bastian's lengthy response to Huxley's speech was published by *Nature* in two parts, on 22nd and 29th of September. that same month.

Bastian accused Huxley of being partial by opposing spontaneous generation to support "atmospheric germ theory":

> *...However reluctant to enter a protest against what has been said by an eminent scientific man, for whom I have always entertained the greatest respect and esteem, I feel so strongly that the representations which have been made concerning a subject to which I have directed the most earnest attention for the last eighteen months, are not only inadequate, but altogether incapable of being regarded as an impartial statement of the main points at issue, that I cannot hesitate as to the propriety of publicly expressing this opinion.*

Then Bastian put forward his findings and justified his accusation with complete confidence. In his efforts to do so, he provided a detailed explanation of his side of the story.

Bastian's first goal, like Haeckel's, was to establish the close resemblance between living and non-living matter and thus end the "special" status given to the living matter. While Haeckel compared the chemical composition of living and non-living matter, Bastian preferred an analogy between the crystalline formation and living matter. Just like the formation of a perfect crystal by the composition of pre-existing matter, living matter is also formed by the composition of different pre-existing matter. So if one ascribes no special property for the crystalline formation, then why attribute the generation of living matter to something special? Bastian would use this analogy frequently in future. He thus refutes any 'vital principle' contributing to the archebiosis.

Once establishing the close analogy between the crystalline formation and the living matter, Bastian deals with the question of whether living matter can arise de novo. An agnostic Huxley had never dismissed the possibility of life stemming from the non-living matter, thus Bastian found no objection from Huxley over such statements. The problem, as per Bastian, was that Huxley thinks there is no evidence whatsoever to prove that life arises de novo. But Bastian had no doubts left regarding the matter:

> *... I affirm that multitudes of minute living things may and do gradually appear in fluids, beneath the microscope, where no visible germs previously existed.*

Bastian then provides the details of the de novo generation of life. The reason why people do not believe in the de novo generation of life is "*in 999 cases out of a thousand which come under their actual notice, there cannot be a question that a living thing originates from a pre-existing living thing*". And when the life generates de novo, they either fail to notice or apply the same rule as when life generates from its parents. Most will fail to notice the life generated de novo because the living thing in the solution could be "*less than 1/100000th of an inch in diameter*". These minute "*visible specks*" of life could have been generated directly by the collocation of the matter in the solution, or they could have grown from smaller germs that are invisible even under a microscope.

Interestingly, Bastian's observation does not stop with the "visible specks" of life. He claims that these life forms grow further into higher life forms like bacteria, and then again into even higher forms –

> *Once formed, it is true, one of these living specks may develop into a Bacterium, and this may develop into a Vibrio or a Leptothrix filament, whilst another of the living specks may develop at once into a Fungus-spore.*

But, Bastian complains, Huxley considers none of these observations and instead recalls the history (of Sphagnum leaf incident). Bastian stated that Huxley's speech mentioned only the people who have contributed towards establishing the existence of atmospheric germs, and even if one would give the benefit of the doubt to Huxley on this point, the experiments conducted proving the archebiosis would still be valid. Bastian confidently informs the reader that the life appears de novo in the solution even after it is subjected to the prolonged heating and sealed hermetically to rule out the possibility of any atmospheric contamination.

Bastian mentioned other evidence he had witnessed, including the processing of canned foods. The food would remain germ-free after going through the rigorous heating process before and after being sealed hermetically. But in some cases, tin cans would suddenly bulge at the corners indicating gas formation inside. And once these cans were opened it was found that the food had undergone putrefaction. This happened even in tins that had no defects. Sometimes the canswould be fine for two or three years and then suddenly they would show this behavior. After verifying those tins himself, Bastian concluded:

> *Certain it is...the circumstance that the failures sometimes take place after the tins have been hermetically sealed for two or three years, and that gelatinous substances are the least prone to change -- are all rather strongly in favour of my view of the case, and will continue to be so, so long as our knowledge concerning the inability of Living things to resist the destructive influence of very high temperatures remains in anything like the same condition as it is at the present day.*

In conclusion, Bastian accused Huxley of being partial, misrepresenting the facts, and calculatedly misleading the public.

Bastian's strong argument against the "omne vivum ex ovo" (all life from life) concept could not be neglected . Huxley had agreed to the possibility of first life originating from the non-life matter, and all X Club members had understood the necessity of abiogenesis for evolution to work. The problem was in accepting the evidence Bastian presented for the spontaneous generation. Huxley and Tyndall could not agree there was and is any evidence available to support the spontaneous generation, but Bastian was confident that he had.

For the first time, Huxley thought of replying to this accusation. So far Huxley had tried to avoid hurting Bastian's feelings, but now Bastian had gone public with serious personal accusations. Huxley's response was published in *Nature* on 13 October 1870,

> *I have been blamed by some of my friends for remaining silent as long as I have done concerning them. But when, because I have preserved a silence, which was the best kindness I could show to Dr. Bastian, he presumes to accuse me publicly of unfairness, and to tell your readers that my Address "is calculated to mislead" them, I have no alternative left but to give them the means of judging of the competency of my assailant.*

Huxley's reply was rather short compared to Bastian's because Huxley found no reason to deal with Bastian's theoretical arguments. He only handled the accusations made against him.

Huxley began by explaining his own efforts in verifying the theory of spontaneous generation. Not a single experiment that was optically pure and free from any particles produced the life spontaneously. Then Huxley examined Bastian's finding of bacteria and Leptothrix in meat preservations that were alive and moving in the boiled solution. Huxley attributed the movement of these germs to Brownian motion. Brownian motion refers to the motion of non-life particles suspended in the liquid or gas due to the collision with molecules in the solution. This is named after Robert Brown (1773-1858), even though this behavior was observed before, as he is widely credited for ruling out the role of "living" particle's motion in Brownian motion. By referring to Brownian motion, Huxley suggested that it was the dead bacteria or Leptothrix moving in Bastian's boiled solution. The preserved meat had dead germs after subjecting to the prolonged heat procedure, and later when Bastian observed the motion, it was nothing but Brownian motion caused by the same dead germs.

Bastian's next complaint was that Huxley seemed to assume there were errors in his works. Huxley brings up example of the Sphagnum leaf fragment once again:

> *Any time these six months Dr. Bastian has known perfectly well that I believe that the organisms which he has got out of his tubes are exactly those he has put into them; that I believe that he has used impure materials, and that what he imagines to have been the gradual development of life and organization in his solution, is the very simple result of the settling together of the solid impurities, which he was not sufficiently careful to see, in their scattered condition when the solutions were made.*
>
> *... Dr. Bastian will remember that when the first of these wonderful specimens was put under my microscope, I told him at once that it was nothing but a fragment of the leaf of the common Bog Moss (Sphagnum); he will recollect that I had to fetch Schact's book "Die Pflanzeuzelle," and show him a figure which fitted very well what we had under the micro-*

> *scope, before I could get him to listen to my suggestion; and that only actual comparison with Sphagnum, after he had left my house, forced him to admit the astounding blunder which he had made.*

Adding to this, Huxley mentions another incident where Bastian had no problems in accepting living things in the potash-alum '*until Dr. Sharpey puts the not unimportant question, whence did they get their nitrogen?*' This would indicate that Bastian was eager to accept the archebiosis without actually subjecting his experiments to a more critical analysis.

Huxley ended his response by addressing only the accusations against him. Bastian did not take it lightly, and he had his own response published in *Nature* within a week on 20 October 1870 in Nature.

While acknowledging the Sphagnum leaf fragment incident, Bastian resisted Huxley for using that one incident to refute many other experiments that clearly show germs being generated spontaneously in hermetically sealed solutions subjected to prolonged heat. As per Bastian, either those germs must have withstood high temperature or generatedspontaneously. And he claimed that the solutions were subjected to such a high degree of temperature that no living organisms should have been able to survive it.

Regarding Huxley's reference to Brownian motion, Bastian asked why he didn't regard those movements as Brownian when he verified the experiments the first time:

> *...Movements which, in the month of March of the present year, Prof. Huxley did not regard as Brownian, he now*

> *does believe to have been of this nature. If he is now right, what value is to be set upon his knowledge of Brownian movements six months ago; and what guarantee have we that in another six months Prof. Huxley may not again take a different view?*

And then he finally concluded by responding to Huxley's comment about maintaining the silence towards Bastian so far:

> *Let me tell Prof. Huxley that I repudiate such "kindness," as any honest man would who is simply seeking after truth, and relegate it to the same regions as I would that indescribable air of restrained omniscience whereby he endeavours to crush arguments and facts, to which he altogether fails to reply.*

After that, the relationship between Huxley and Bastian turned sour. Huxley did not bother responding to Bastian, and Bastian became consumed with writing a book on the same subject.

A Warm Little Pond

Meanwhile, all these developments made Darwin reconsider his position about the origin of life and the possibility of abiogenesis. Initially, he was open to accept the possibility, but later grew skeptical. He considered it highly likely that man will never know the answer for the origin of life with complete certainty.

After Bastian's initial three-part article in response to Tyndall's presentation, Darwin had expressed doubt that the germs might have survived the high temperature. Now he had come across a study by Benjamin Thompson Lowne published in January 1871, which referred to the mold spores that could survive prolonged boiling. This was the first study that had contradicted all the previous studies like the Wyman's. In his letter to Hooker on 1 February 1871, Darwin says:

> *It will be a curious discovery if Mr. Lowne's observation that boiling does not kill certain moulds is proved true; but then how on earth is the absence of all living things in Pasteur's experiment to be accounted for?*

This question by Darwin was highly relevant. Once again he had expressed doubt that the organisms could have survived the high temperature. Pasteur had cherry-picked his own results by ignoring the experiments that had come up with the germs. Pasteur could have thought that he inadvertently let the solution contaminated in those negative results, but could it be possible that the germs actually survived the boiling he subjected them to?

In the same letter Darwin mentioned his hope for the abiogenesis:

> *It is often said that all the conditions for the first production of a living organism are now present, which could ever have been present.— But if (& oh what a big if) we could*

conceive in some warm little pond with all sorts of ammonia & phosphoric salts,—light, heat, electricity &c present, that a protein compound was chemically formed, ready to undergo still more complex changes, at the present day such matter would be instantly devoured, or absorbed, which would not have been the case before living creatures were formed.

Darwin's particular view would become famous in the history of the origin of life. Darwin's conception of this idea of the origin of life differed from what was proposed by Bastian. Darwin was suggesting that abiogenesis occurred only once at the beginning when there was no life. But Bastian had taken a very big leap by claiming that abiogenesis happens all the time. Of course, for Bastian, he had experimental evidence and Darwin's notion was just a speculation, albeit a wonderful one.

The Beginnings of Life

A year later, in 1872, Bastian published *The Beginnings of Life: being some account of nature, modes of origin and transformation of lower organisms.* The book followed a similar line of argument as presented in his reply to Huxley, but with a massive amount of details of the supporting evidence.

Bastian started his book by attacking the vital force theory to remove the barrier between organic and inorganic matter. He then fiercely attacked Pasteur's experiments. He accused Pasteur of making too many assumptions to fit his views without sufficient evidence. Bastian suggested that there was no evidence to prove the existence of germs in the atmosphere. Even if the atmospheric germs theory were true, it would make no difference because Bastian had performed the experiments avoiding any atmospheric influence and yet the germs were generated spontaneously. Also, according to Bastian, Pasteur concentrated only on the negative (no germs generated spontaneously) results and thus completely ignored the positive results. Moreover, Bastian had found more positive cases than negative after following Pasteur's methods. So, Bastian stated, almost every experimenter, including Pasteur, who had rejected spontaneous generation could only do it based on some theoretical explanation rather than experimental evidence. As per Bastian, if one could spend just three months of time on the experimental observations, then he cannot reject the evidence he would witness.

Later in his book, Bastian takes on the atmospheric germ theory by providing instances where the germs suggested to be in the atmosphere were found inside the bodies of higher organisms. For example, some fungi and bacteria were generated inside living plant or animal cells, and there was no way those organisms could have entered the cells through the cell walls. Hence, Bastian says, these cases observed even by many other biologists are evidence for heterogenesis.

Bastian rules out the role of the atmospheric germs based on the experimental outcome. He had sealed the flask while boiling the solution, and then he subjected it to a greater degree of heat for a prolonged time. It was known at the time that the fungi spores survive greater degree of heat than any other germs, but all the studies so far had shown that in the solutions heating at 212^0 F for few minutes or dry heating outside the solutions at 266^{0} F for about 30 minutes would kill all the germs. But Bastian had subjected his solutions to a much greater degree of heat, more than 275^{0} F, for about thirty minutes and still found the life originating afterward. Bastian named many biologists like Wyman, Mantegazza, and Cantoni who had conducted similar experiments and came up with similar results. And he does not deny that in some cases no life was originated. But those few cases do not prove that archebiosis does not happen. Instead, they prove that the favorable conditions for spontaneous generation to occur including the room temperature, composition of the solution, quantity, and many other factors are interrelated in a very complex manner. It would require an in-depth study and multitude of experiments to understand the true nature of these conditions.

Bastian's experiments supported Pouchet's observations. The experiments had produced minute visible living matter that was similar to what Haeckel had described as Monera. Bastian had even found the organisms with a very close resemblance to Haeckel's Protista. Kingdom Protista was proposed by Haeckel to keep the lower forms of life separate from plant or animals. Bastian observed that these lower life forms were going through the heterogenesis process with great ease and hence producing many new life forms. For the same reason, these life forms cannot be grouped into any specific species. So Bastian introduced a new term for them – Ephemeromorphs:

> *...these lower forms of life—corresponding pretty closely with the Protista of Prof. Haeckel—form an enormous and*

ever-growing plexus of vegetal and animal organisms, amongst which transitions from the one to the other mode of growth take place with the greatest ease and frequency. Here Heterogenesis is constantly encountered, and variability reigns supreme, so that those assemblages of definitely recurring individuals, answering to what we call "species" are not to be found amongst them. They are essentially transitory and variable forms, which I have proposed to name "Ephemeromorphs." Regularly recurring or homo-genetically produced types, both animal and vegetal, are, however, constantly arising out of this great ephemeromorphic plexus, either by direct and sudden processes of transformation or by some intermediate and cyclical processes of so-called alternategeneration.

After claiming with absolute certainty that he had the evidence for the archebiosis, Bastian moves on to heterogenesis. As per Bastian, the first visible specks of life came from the invisible living matter, which in turn are generated by non-living matter. Once these minute visible organisms appeared, they grew further and transformed intocompletely different organisms. He claimed that he had observed these stages of the transformation very closely in a multitude of experiments and described them in detail. Sometimes the minute visible specks grew into ameba and then into bacteria, and in some other cases they would develop to become fungi. The question was why they would grow to form different organisms in different cases. Bastian acknowledge that he had no definitive answer to this question, but theorized that the difference depended on a number of factors such as the constituents forming the solution, their quantity, and the temperature used in the experiment.

Theoretically, Bastian argued that if one assumes that spontaneous generation does not generate the smallest organisms like the bacteria, then that conclusion infers that all microorganisms have been propagated through reproduction for millenia without undergoing any evolution. That would seem improbable and absurd. And on the other hand, if they generate spontaneously, it would explain why time and again we notice those same organisms.

But Darwin had answered to a similar type of argument in his response to Richard Owen's anonymous review of Dr. Carpenter's book in the *Athenaeum* in 1863. One complaint from Dr. Carpenter was that the foraminifera had not undergone many changes since ancient times, which would be an objection to evolution theory. Darwin had replied that it was the a Lamarckian assumption that the organisms always evolve to become better. Evolution, as per Darwin, is of the variations retained when found to be favorable to the organism's survival, and this variation may not necessarily make the organism more complex or even "better." This is why it is difficult to say why one group of organisms undergo rapid evolution but not the other. As Darwin wrote:

> *...lowly-organized animals are best fitted for humble places in the economy of nature; that an infusorial animalcule or an intestinal worm, for instance, would not be benefited by acquiring a highly complex structure. Therefore, it does not seem to me an objection of any force that certain groups of animals, such as the Foraminifera, have not ad-*

vanced in organization. Why certain whole classes, or certain numbers of a class, have advanced and others have not, we cannot even conjecture. But as we do not know under what forms or how life originated in this world, it would be rash to assert that even such lowly endowed animals as the Foraminifera, with their beautiful shells as figured by Dr. Carpenter, have not in any degree advanced in organization. So little do we know of the conditions of life all around us, that we cannot say why one native weed or insect swarms in numbers, and another closely allied weed or insect is rare. Is it then possible that we should understand why one group of beings has risen in the scale of life during the long lapse of time, and another group has remained stationary?

Support for Bastian

Huxley, Tyndall, and rest of the X Club members were not impressed by Bastian's book. As long as Bastian's experiments were not proved, they had no reason to be. But the younger evolution supporters were rather confused by all the developments. This younger generation would look to senior scientists for advice whenever needed, and initiallyBastian had been one of the strongest among this younger group. But Bastian's firm stance against the senior members caused a lot of confusion and disappointment. Bastian was not alone in his fight. He had support from many of his colleagues and friends. Moreover, like adding more fuel to the fire, Alfred Russel Wallace vouched for Bastian in a very positive tone.

Unlike Darwin, Wallace was very excited about Bastian's

work. Wallace held a positive view about Bastian's experiments even before *The Beginnings of Life*. He wrote a lengthy review of the book that *Nature* published in two parts in August 1872. In his complete review, Wallace finds absolutely no concern over Bastian's views except for his opinion on vitalism. Wallace argues that the vital force could still exist, like the gravitational force or electricity, and play some role in the generation of those incredible minute organisms. Besides this one minor point, Wallace praises Bastian's enormous experimental evidence and arguments throughout. Wallace says that Bastian's carefully devised experiments not only answer all the objections to his previous works such as Huxley's mention of the Brownian motion but also shows the "*fallacy of many of the arguments of M. Pasteur and his supporters*".

Wallace mentioned that the findings of Bastian were in perfect line with Darwin's theory. In fact, he ventured to say that it resolved many difficulties surrounding the origin of life that Darwin had trouble explaining. For example, Darwin held that the early life forms, being simple with slow reaction times, must have gone through a very slow process of evolution. But at the time the age of the Earth was thought to be only a few thousands of years, so one of the strong objections to Darwin's theory was that geologically it would be impossible for evolution to produce such diversity in a relatively short amount of time. But if Bastian is right, Wallace says, then this objection no longer holds. The minute life forms are undergoing a rapid and frequent transition though heterogenesis, which would lead to complex organisms in much shorter time. Proceeding further, Wallace adds that there is not a single tree of evolution but rather many that have originated from low forms of lives at different times.

Wallace sought Darwin's opinion on the same, but as Darwin replied on 28 August 1872 he was not very impressed with Bastian's book: "*You wished to hear my impression, but it is not worth sending.*" Darwin agrees that Bastian's argument for archebiosis is strong, but at the same time the experiments failed to convince him. For example, regarding the Bastian's heat experiments Darwin says –

> *I must have more evidence that germs, or the minutest fragments of the lowest forms, are always killed by 212 degrees of Fahr. Perhaps the mere reiteration of the statements given by Dr. Bastian [by] other men, whose judgment I respect, and who have worked long on the lower organisms, would suffice to convince me. Here is a fineconfession of intellectual weakness; but what an inexplicable frame of mind is that of belief!*

Darwin had reservations because Bastian's experimental procedures had been questioned by X Club members that Darwin respected. Darwin himself had faced harsh criticism for the theory of evolution even after the decades of work collecting a multitude of evidence. So not surprisingly, he was not in a hurry to trust Bastian's experiments based on nothing but Bastian's own words.

Darwin was also skeptic about the maximum temperature that the consensus believed would be good enough to kill any living organism. Further, Darwin could not agree with Bastian's claims on spontaneous generation of some of the complex organisms:

> *As for Rotifers and Tardigrades being spontaneously generated, my mind can no more digest such statements, whether true or false, than my stomach can digest a lump of lead. Dr. Bastian is always comparing Archebiosis, as well as growth, to crystallisation; but, on this view, a Rotifer or Tardigrade is adapted to its humble conditions of life by a happy accident, and this I cannot believe...He must have worked with very impure materials in some cases, as plenty of organisms appeared in a saline solution not containing an atom of nitrogen.*

But Darwin was not opposed to archebiosis itself. To the contrary, he was very much for it. He confirmed to Wallace that "*it seems to me probable that Archebiosis is true.*" So Darwin mentions that:

> *I should like to live to see Archebiosis proved true, for it would be a discovery of transcendent importance; or, if false, I should like to see it disproved, and the facts otherwise explained; but I shall not live to see all this. If ever proved, Dr. Bastian will have taken a prominent part in the work.*

Wallace, knowing Darwin's struggle against all the attacks on his theory, understood his stance but he also sensed that the Sphagnum leaf incident could have had some effect on Darwin. He addressed it in his response on 31 August 1872:

> *I quite understand your frame of mind, and think it quite a natural and proper one. You had hard work to hammer your views into people's heads at first, and if Bastian's theory is true he will have still harder work, because the facts he appeals to are themselves so difficult to establish. Are not you mistaken about the Sphagnum? As I remember it, Huxley detected a fragment of Sphagnum leaf in the same solution in which a fungoid growth had been developed. Bastian mistook the Sphagnum also for a vegetable growth, and on account of this ignorance of the character of Sphagnum, and its presence in the solution, Huxley rejected somewhat contemptuously (and I think very illogically) all Bastian's observations. Again, as to the saline solution without nitrogen, would not the air supply what was required?*

Wallace's point was that it is illogical to discard all the observations based on that one incident. But the whole situation was more complex than what Wallace was suggesting. Tyndall's impressive experiments observing the particles in the air through a beam of light were far more accurate and unquestionably error-free. Huxley and Tyndall had repeated most of the experiments done by Pasteur, Schwann, and predecessors and came up with same results. Even in case of Bastian's experiments, in optically pure solutions they had never obtained the results that Bastian had claimed.

But the whole heated situation confused some of the young evolutionists. Some supported Bastian, and many others, like Anton Dohrn, were disappointed with him. Dohrn a good friend of Huxley, sent a letter to Darwin on 21 August 1872 soon after reading Wallace's review in *Nature*:

Poor Wallace completely drifts away... His two articles in Nature are the worst thing he ever did in his life, and it becomes really difficult for his friends to speak with respect of him.

Pasteur vs. Bastian

Bastian continued to attack Pasteur and the atmospheric germ theory by conducting more and more experiments. *Comptes Rendus*, a French science journal, published his works on the fermentation in 1876. In his paper, Bastian presented evidence showing that the germs in fermentation are not atmospheric germs as claimed by Pasteur but instead were generated spontaneously in the solution. Pasteur was a household name in France at the time. He held a patent for the process of pasteurization, handled the chicken cholera, and was now working on a vaccine for anthrax. So far he had ignored Bastian, but when *Comptes Rendus* published the paper on the fermentation, he thought of responding. He invested little time in preparing the reply because he had already applied the germ theory to practical use many times over. He had come so far in such applications that a question on this theory seemed meaningless. So he was convinced that Bastian was wrong.

Bastian had experimented on urine. He had taken boiled urine, saturated it with potash, subjected it to a heat of 50^0C, and still found that bacteria generated spontaneously within the solution. Pasteur replied to Bastian's paper in *Comptes Rendus* on 17th July, reminding Bastian of his 1862 paper on the organic corpuscles in the atmosphere in which he had conducted a similar experiment. Pasteur had shown in that article that the acid solution that would become sterile at 100^0C would not be sterile if it had somehow become slightly alkaline. There were minute organisms that would survive at 100^0 C in alkaline solutions, and Bastian was trying to sterilize a neutral or slightly alkaline solution at 50^0 C. So this would only prove there are minute organisms that can survive that temperature. The problem, Pasteur says, is with the liquid potash Bastian adds to urine solution, which is impure. Instead, he suggested using solid potash and heating it to red-hot or at 110^0 C. Pasteur asserts that Bastian will never find any organisms if he employs this process and mentions "*I entertain the firm hope that he will abandon his belief in spontaneous generation and in the proofs he supposes he has adduced.*"

But Bastian was not in a position to doubt the results of his experiments. This time Pasteur had no intention of immersing himself in the same experiments again to try disproving Bastian. For him, Bastian was just like Pouchet, a subpar experimenter who jumps to conclusions too quickly without analyzing the flaws in his method.

Tyndallization

As all these events were happening, Tyndall was silently working in his Royal Institution laboratory to put a full stop to Bastian's efforts. As Pasteur had prepared himself for the debate with Pouchet, Tyndall too prepared himself thoroughly before attacking the theories of Bastian and his supporters.

Tyndall began by conducting an experiment in Pasteur's

style. He prepared the sixty hermetically sealed flasks containing the boiled and sterile infusions of beef, mutton, turnip, and cucumber. He took them to the Alps, about 7000 feet above sea level, and opened them. He noticed that fifty-four of the flasks remained pure with no microbial growth observed, whereas the other six remaining flasks were broken during the transit to the Alps and hence contaminated. Then he sealed those fifty-four flasks containing the pure infusions and kept them in warm weather, a condition considered favorable for the putrefaction, but even then there was no spontaneous generation. To answer a possible objection that those flasks needed air for the spontaneous generation to occur, Tyndall divided the flasks into two groups of twenty-seven and twenty-three, as four flasks were broken again during the time. He took twenty-three of the flasks to a hayloft and carefully opened them. All these flasks, except for two, were swarmed with minute organisms within three days. Then he took the remaining twenty-seven flasks from the second group and opened them from a place 200 feet above hayloft. He observed that even after three weeks, no flasks showed any microbial growth. Tyndall asks:

> *Is not the inference here imperative that it is not the air of the loft—which is connected through a constantlyopen door with the general atmosphere—but something contained in the air, that has produced the effects observed? What is this something? A sunbeam entering through a chink in the roof or wall, and traversing the air of the loft, would show it to be laden with suspended dust particles.*

Tyndall then picked another claim that said "*withdrawal of the atmospheric pressure above the infusion favours the production of organisms.*" This was explained particularly in reference to the preserved food in the tins. About the tins,the claim was that the fermentation process begins after the package is done and that leads to the production of gases. These gases in a closed container create atmospheric pressure blocking the further generation of organisms. Once the tins have opened the gases, atmospheric pressure will be released, leading the way for the growth and generation of organisms. But, Tyndall says, if we open the tins under the water, we can clearly notice that the water enters the tin, but gases won't come out of it. Tyndall repeated this process using a sixty-three-year-old tin in the Royal Institution as well as new tins and finds that none of the tins with preserved food had any gas in them. Further, he conducted another experiment by subjecting the infusions in the iron bottles with an air pressure of 10-27 atmosphere pressures and found that after ten days the infusions had still undergone putrefaction. This, Tyndall says, proves that the removal of atmospheric pressure from the gases has no role in the spontaneous generation.

Next Tyndall came back to the experiment that made him famous: the phenomena known as "Tyndall Effect" where the particles suspended in the air cause light to scatter. When he had published his results confirming the experiments of Schwann or Helmholtz using a strong light beam to confirm if any floating matter exists in the air, Bastian had asked him not to indulge in questions of biology and instead stick to physics. Now, Tyndall had improved this experiment further.

To demonstrate his findings, Tyndall first placed a spirit

lamp near the beam of light. He noticed there was darkness near the rims of the flame as if it was smoke. But he was using the spirit lamp, so that coundn't be smoke. To confirm, he repeated the same by using a hydrogen flame that also does not emit smoke. Once again he noticed the same. Upon analyzing he figured that the flame destroyed the floating matter surrounding it, and the heated air now moved upwards. Considering that the floating matter scatters the light, it would give an illusion of smoke. Now, if he placed infusions in the particle-free and calcinated air, they would not undergo the putrefaction process no matter how long he waited.

But one might still object that the air is "spoiled" by the flame used. So Tyndall built a different setup to conduct this experiment. In his first and most famous experiment, Tyndall constructed a wooden chamber with a glass wall on four

John Tyndall's setup to free the subject from floating particles

sides of it. A provision was made for the test tubes to enter the chamber from the bottom, and there was a circular hole onthe top through which the pipette would make way. There was also a provision for outside air to enter the chamber. All these were carefully designed to be airtight, so no contamination was possible.

The chamber was closed airtight once the outside air entered and was then left in that state for three days. Tyndall had applied glycerin to the inside of the chamber walls so any floating particle that touched the wall would stick to it. That way, after three days chamber was completely free of any floating particles. Now the six test tubes containing the solutions were added to the chamber. Before placing them in the chamber, the test tubes were immersed in heated oil bath,and the solutions were boiled to achieve sterility. No matter how long these tubes were kept inside the chamber devoid of any floating particles, he observed neither the putrefaction nor the generation of new life. Tyndall repeated this experiment using more than fifty chambers with different solutions for more than a year and not a single time did he witness spontaneous generation.

He just needed to allow the air inside those chambers to confirm there was "something" in the air that causes the putrefaction. Only within three days of letting the air inside, he noticed that the solutions were full of infusoria. Tyndall now tells Bastian, whom he preferred to refer as "colleague" or "heterogenist" rather than taking his name:

> *He must, I submit, bow to the conclusion that the dust-particles are the cause of putrefactive life. And unless he ac-*

cepts the hypothesis that these particles, being dead in the air, are in the liquid miraculously kindled into living things, he must conclude that the life we have observed springs from germs or organisms diffused through the atmosphere.

But Tyndall knew these experiments would not stop Bastian or his supporters. Bastian had claimed that he had repeated the experiments of Pasteur and others and still obtained results—"positive" results as he called them—where the life appeared spontaneously. Bastian had published such results in his book and argued that we cannot ignore these "positive" cases as Pasteur had done. That was the most difficult challenge ahead of Tyndall. He took a long timeanalyzing Bastian's experiments until he hit upon the answer.

Bastian had claimed that he had heated the solutions at a temperature higher than the maximum limit that all the studies, including Pasteur's, had shown to be fatal for all life. But could germs be surviving that high degree of temperature? And if they don't, how did Pasteur's experiments not produce any germs as Bastian's did? Darwin had expressed this concern few times in his letters. And Tyndall had just found the answer!

It was a well-established fact that different seeds have different resistance limit to heat, and Tyndall had discoveredthat the spores of "hay bacillus" or "bacillus subtilis" bacteria could survive very long exposure to high temperature, as long as more than 8 hours of boiling:

I do affirm, not as a 'supposition,' nor an 'assumption,' nor a 'probable guess,' nor as 'a wild hypothesis,' but as a matter of the most undoubted fact, that the spores of the hay bacillus, when thoroughly desiccated by age, have withstood the ordeal mentioned.

Tyndall also figured an effective way to destroy such seeds or spores through a process known as “Tyndallization“ or “discontinuous heating“ process. The solution should be brought near its boiling point after ensuring that no more external contamination is possible. At that point, all the spores that have reached the final stage of their development will be destroyed. Then keep aside the solution for about ten to twelve hours by which time the spores that survived the previous boiling would have developed further. Then once again subject the solution to heat near the boiling point, which kills the spores then reaching their final stage of development. This process should be repeated few times to ensure that all the spores are destroyed. This “discontinuous heating” process ensures that the solution used in the experiments is completely sterile. Tyndall used this method on the solutions containing hay bacillus and other spores and confirmed that none survived the process.

Tyndall’s “discontinuous heating“ process would solve most of the conflicting results observed by the supporters of spontaneous generation. But he wanted to handle all the specific strong cases favoring the theory. So he picked another claim by the heterogenists of the time. Expose a “mineral” solution containing all the ingredients necessary for the life to the air for an extended time. And it was found that no life appeared in the solution no matter how long it was kept that way. But infusoria appeared in the turnip infusion when exposed to the air in the same atmosphere. This, Bastian and supporters claimed, proved that there are no germs in the air because in that case life would have appeared in both the solutions.

Tyndall said that they had drawn the wrong inference from their observations. To prove it, he placed both the solutions in an airtight chamber with no possible contamination and added drops of the boiled mutton meat juice full of infusoria to both the solutions. Within next day he found that both mineral solution and the turnip infusion had infusoria swarming. So it was clear that both the solutions were capable of providing the necessary nutrition for the survival and growth of infusoria. As a next step, he took another set of both clear solutions into a different chamber and dropped dust into the chamber. So now the chamber's air was contaminated with germs. But he found that only the turnip infusion developed the signs of infusoria within three days. The mineral solution was as clear as it was initially. No matter how often one repeated the experiment, the result was same. So, Tyndall says, the correct inference to be drawn from the experiment is that the mineral solution is incapable of providing the necessary environment for the germs in the atmosphere, but turnip infusion is very much capable of it. He asks Bastian and his supporters:

> *Why, when your sterilized organic infusion is exposed to optically pure air, should this generation of life de novo utterly cease? Why should I be able to preserve my turnip-juice side by side with your saline solution for the three hundred and sixty-five days of the year, in free connexion with the gen-*

> *eral atmosphere, on the sole condition that the portion of that atmosphere in contact with the juice shall be visibly free from floating dust, while three days 'exposure to that dust fills it with bacteria?' over-sanguine in hoping that as regards the argument here set forth he who runs may read, and he who reads may understand?*

Finally, Tyndall attacked Bastian's claim that the "briefest exposure" to the boiling water is destructive for all higher or lower organization of life. Applying the effect of the boiling water on higher forms of life to the atmospheric germs is erroneous. The temperature needed to kill the bacteria in an infusion heated in a contamination-free environment is called the "death-point" of bacteria. But this death-point temperature is not enough to kill the floating "germinal matter" in the air. In fact, Tyndall says, the term death-point is very vague. For example, if an infusion takes 5 hours of boiling temperature to become sterile, the same infusion might take five days to become sterile if the temperature was less than the boiling point.

On the subject of death-point temperature, Tyndall experimented using the turnip infusion subjected to the atmospheric hay germs. He subjected ten flasks filled with the turnip infusion to different intervals of boiling heat, ranging from 15 minutes to 240 minutes. The temperature used was same for all the infusions, so the only difference was in the duration of boiling. Only the one heated for 240 minutes was sterile, whereas all others had infusoria. Tyndall thus concludes –

> *...we have gone far beyond the argument, and proved by*

multiplied experiment the alleged destruction of all living matter by the briefest exposure to the influence of boiling water to be a delusion.

Tyndall published his findings in 1876. Tyndall had not pushed back against Bastian's criticism until he knew hisarguments were too strong to contend. His "Tyndallization," or "discontinuous heating," method had finally delivered the deathblow to the theory of spontaneous generation. Though Bastian was still unconvinced by Tyndall's findings, he went silent after that, and so did all the other spontaneous generation supporters.

The End of an Era

Darwin died on 19 April 1882. Throughout his life, he had opposed applying his theory to the society, and yet social Darwinism and eugenics were unquestionably on the rise. Darwin's half-cousin Francis Galton (1822–1911) coined the term "eugenics" in 1883 and famously used the phrase "nature vs. nurture". He suggested that offspring would get better hereditary traits through controlled breeding in humans. Eugenics became very popular in many countries, and some countries even passed laws to implement this idea for a while.

Tyndall retired from the Royal Institution in 1887 due to illness. He suffered from severe insomnia and had to take chloral hydrate as part of the treatment. His wife Louisa Hamilton administered the drug daily, and on 4 December 1893,she overdosed him with the chloral hydrate by mistake.

Even though the doctors tried to counter the damage, that mistakeclaimed Tyndall's life. His guilt-ridden young wife invested herself in writing Tyndall's biography, which was published five years after her death in 1940.

Huxley retired from work couple of years earlier than Tyndall did, in 1885. His family, including his father, brother, daughter had all suffered from depression or anxiety disorders. Huxley had also gone through occasional episodes of depression. And when it returned in severe form once again in 1884 he decided to retire. In 1895, after a bout of influenza and pneumonia, he died of a heart attack on 29 June 1895.

Beginning with Huxley and Tyndall's illnesses in the 1880s, the X Club had started falling apart. William Spottiswoode succumbed to typhoid fever in 1883. George Busk died two days short his 79th birthday in 1886. In 1889, Huxley and Spencer had a fight over a political subject and stopped talking to each other. Thus Spencer moved out of x-club. In 1892, Tyndall's friend Thomas Hirst died. With only a few surviving members, they met less frequently. They could not come to terms about choosing new members of the club, so finally in March 1893, the last meeting was held with only Hooker and Frankland participating.

That was the end of the famous X Club.

One Last Futile Attempt

Suppressed by the overwhelming opposition from different quarters, including Tyndall's masterstroke, Bastian retired from his quest to prove the existence of archebiosis and heterogenesis after 1876. He published important works on paralysis, epilepsy, and hysteria. His book *The Brain as Organ of the Mind* in 1888 was one of his most important works published. He became one of the most renowned pioneers in his field as a neurologist.

Nevertheless, Bastian was disappointed that nobody was willing to verify his experiments on spontaneous generation. Instead, the scientific community unquestioningly accepted Tyndall's experiments and counterattacks. He had no choice but to go dormant for a while. But once all his opponents had passed away, he bounced back with the same conviction, if not more, and started conducting experiments related to archebiosis and heterogenesis with renewed vigor.

From 1900 until his retirement from the National Hospital in 1912, Bastian published four books on the same subject—*Studies in Heterogenesis* (1904), *The Nature and Origin of Living Matter* (1905), *The Evolution of Life* (1907), and *Remarks on Further Experiments Concerning the Origin of Life* (1912). He followed the same theoretical argument but varied only his experimental methods.

Bastian failed to realize the huge advancements in the germ and sepsis theories, and that medical authorities had universally accepted them. By that time, people had much better understanding of diseases, the way they spread, and the methods to contain them. With all these developments happening, the idea of spontaneous generation seemed old and outdated. People simply stopped taking the spontaneous generation seriously. While the question of the origin of life was still left wide open for studies, it was clear that spontaneous generation was no longer a favorable option. As a result, to Bastian's utter dismay, nobody had the patience to examine or even consider his experiments. This made him the last scientist who supported the theory of spontaneous generation.

While Bastian invested much of his time and energy on the subject of the origin of life, other eminent neurologists of his day such as John Hughlings Jackson (1835–1911) and Sir James Crichton-Browne (1840–1938) surpassed him in neurology. Even though Bastian's contributions to his field were significant and remarkable, his contemporaries later overshadowed his name.

Without a doubt, even among all the controversies surrounding his debates on the origin of life, Bastian had an excellent career and reputation. But it is a sad fact that he had to struggle financially during the last years of his life. Hehad no money saved for his retirement in 1912, and he received no pension from the university or the hospital. His booksfailed commercially, and he had to rent a small home due to the financial constraints. Finally, his former colleagues and well-wishers like Sir James Crichton-Browne, Sir James Geikie, Sir Thomas Barlow and Sir William Ramsay wrote to then Prime Minister Herbert Henry Asquith and made sure that he received a pension of 150 pounds per year in 1914. But Bastian did not live long after that. A year later, on 17 November 1915, he breathed his last.

With Bastian, the millennia-old theory of spontaneous generation had died.

CHAPTER 6

LIFE FROM SPACE

Most of the efforts at finding an answer for the origin of life until the end of 19th century were directed at the mechanism of origin on Earth. But considering we live in such a vast universe, is it not possible that life began outside our small planet? The term "life" does not necessarily need to refer to fully functioning, compound living organisms. It might refer to the seeds of life, or rather, building blocks of life. What if these building blocks of life exist in space? What if they might have reached Earth in some way and led to the origin of all life on this planet?

A small but significant number of philosophers and scientists have put forth theories suggesting the same. It started as a purely philosophical idea in ancient Greece and transformed into a full-fledged scientific hypothesis by the end of 19th century. We see the first account of such a suggestion in the works of Anaxagoras. The theory of "everything in everything" proposed by Anaxagoras mentions that everything came forth from a primitive mixture that included innumerable seeds of the life. These seeds of life were the source of the life on Earth. As the primitive mixture and the seeds of life within it existed outside the Earth, it implies that the source of life on Earth was external to it. Though we don't know how Anaxagoras came to such conclusion, he gives us an idea that space contains seeds of life within it, and by stretching this idea a little further we can make an assumption that life reaches earth from outside space.

Telliamed

We derive the idea of life from space from the works of Anaxagoras, but it took another two millennia to see a similar proposal in the works of Maillet. Benoît de Maillet (1656–1738) was a French diplomat who traveled extensively for his job. An amateur geologist as well, he studied fossils from mountains and sea beds and concluded that the Earth was once covered entirely with water. The water level reduced over time, but at a steady rate. He calculated Earth's age to be around 2.4 billion years based on the height of the highest mountain he could find from sea level and considering a value he derived for the rate at which water level reduces on Earth. Further, his study of the Earth's crust led him to conclude that Earth did not form rapidly, but rather slowly and gradually over millions of years.

As per Maillet's calculations, water covered the Earth completely for about 2 billion years. Considering that there was water everywhere for such a long period, his theory for the origin of life suggests that the life appeared first in the sea. Later, following Maillet's version of the evolution theory, life moved to the land. But how did life first appear in the sea? Was it generated spontaneously?

The idea of Earth being 2.4 billion years old was itself revolutionary in the 18th century. But De Maillet went even further than his peers and attempted to explain the origin of life by suggesting that the seeds of life are everywhere in the universe.

In 1732, Maillet gave a collection of manuscripts written over a decade to Abbott Jean Baptiste de Mascrier for editing. He called it *Telliamed*, which is his own name written in reverse order. He placed his arguments in the form of the dialogue between an Indian philosopher and a French missionary. He took the role of the Indian philosopher and presented all the theories through him.

According to Maillet, the seeds of life from space initially entered the water that had covered the Earth. This led to different life forms in the sea. Sea life evolved when the continents started appearing, and various organisms moved to the land. Trees, birds, and man all came from the sea in different forms. In the book, Maillet starts with the origin of the system, the mechanics of its formation, and provides evidence in support of his theory.

In every male, Maillet says, there is a first uterus used for the fertility of the seeds. These seeds of life are distributed across the universe and are numerous in the atmosphere and in the water, but less numerous between the planets because space lacks the arrangement to keep them fixed. Every living thing consumes these seeds when it respires. The law of nature says that everything should be attached to its own species, and following this law, the seeds belonging to the specific species will reunite in the first uterus once the male reaches the puberty. When they reach the second uterus in the female, they acquire more strength to grow and are finally pushed out into the water or air, based on the nature of the species.

When there was still no life on the Earth, for the first time, seeds from space entered the water and followed a similar approach at the bottom of the sea. In this case, the first uterus was the water itself and the second uterus was any suitable matter like the slime that could provide much-needed heat. At a particular time and disposition, the waters reached a stage, similar to puberty, that was suitable for the growth. Later, the seeds found an appropriate second uterus, like slime, depending on the diversity of the dispositions available at the bottom of the water where they fully grew. And different forms of life developed depending on the diversity of the disposition.

Maillet further discussed the microscopic observations of these seeds. Before finding a suitable uterus, the seeds are so small as to be invisible even under the best microscope. But if we take a drop of water from the vessel, we can see a huge number of animals belonging to different species, including man. This proves the possibility that all animals, in the air or on the land, came from the water. And, as these animals in the drop of water are too small initially, we won't notice the body parts. But if we keep the observations going we can see the organs appearing day by day as they keep growing. Here, Maillet is suggesting preformationism.

The theory of the seeds of life is discussed at the end of *Telliamed*. Before reaching this point, Maillet explains how different forms of life have appeared and how every life form has evolved from the sea life. One can witness the sea life resembling all the life forms that now exist, such as lions, wolves, camels, cats, and horses. If animals came out of the sea, even birds evolved from the flying fishes. It is easier to accept this transformation of birds when we compare it with the metamorphosis of a silkworm or butterfly. After dealing with all the animals, the book comes to the question of the origin of man. Maillet rejects all the old ideas as the works of fancy and imagination, and offers his own explanation of how man came from water. He cites events from the different places like Egypt, Netherlands, and Arabia where many have supposedly witnessed the sea-man or sea-woman. Maillet considered these to be the evidence for man coming from the water.

Maillet acknowledges that his theory could be wrong, but he refuses to believe the ancient theories that were still popular during his time. His view in this regard is summarized in this passage from *Telliamed*:

> *To say that our ancestors believed a thing, is a pitiful and pernicious argument, which cramps the Mind, destroys reason, favors Ignorance and Error, and only proves, that in all Ages Men have been credulous; that since the number of ignorant and foolish Men is greater than that of the wise and sagacious, truth is not always to be eliminated by the multitude; that the more ancient an opinion is the nearer it approaches to the fabulous times; and that consequently there is no sentiment worthy of reception, than that which has no more solid foundations than those of time, and the multitude of the votaries.*

Though Maillet finished *Telliamed* by 1732, it was not published until 1748, ten years after his death. He trusted his editor Mascrier to edit and publish the book. Seeing the potential hullabaloo upon publishing it, Mascrier took his time in editing to make it sound less controversial and to bring it in line with Catholic ideas as much as he possibly could. But the very nature of the book would make it impossible to remove all controversial ideas.

For Voltaire, Maillet belonged to the same class as Needham or Buffon – a threat to religion. He attacked Needham in the "Story of the Eels on Which the System is Founded" and then proceeded to deal with Maillet:

> *...it is yet more deplorable that natural philosophers, of better information, adopted the Jesuit Needham's ridiculous system, and joined it to that of Maillet, who asserted that the ocean had formed the Alps and Pyrenees, and that men were originally porpoises, whose forked tails changed in the course of time into thighs and legs. Such fancies are worthy to be placed with the eels formed by meal.*

Voltaire wrote the "The Man of Forty Crowns" in a similar style to *Telliamed*. Maillet had used the conversation between an old hermit and the Indian philosopher, with the Indian philosopher placing the theory and arguments. In this case, Voltaire used the old hermit to place counter arguments for all the points raised by the Indian philosopher:

> *Who taught me that mountains and men are made by the sea waters. First, there were handsome mermen that later became amphibians. Their beautiful forked tail transformed itself into buttocks and leg. I was full of Ovid's Metamorphoses, and a book where it was shown that the human race was a bastard of a race of baboons. I liked descending from a fish about as much as from a monkey. With time I*

had few doubts about this genealogy, and even about the formation about mountains.

Maillet's seeds of life theory gained no support and was soon forgotten. Nor did it change the mood of the scientific community regarding the subject of spontaneous generation. The majority consensus lined up against the Maillet's theory because of its revolutionary nature and the total lack of evidence. As Voltaire pointed out, it sounded completely illogical to think that man evolved from sea life.

Organic Matter in Space

Anaxagoras and Maillet's accounts of life from space were philosophical in nature. Neither man submitted evidence to support their claims. But some interesting developments hinting at the possibility of life in space were seen at the beginning of 19th century.

Meteorites

Berzelius's significant contributions in bringing vitalism to prominence in the early 19th century are well known. Even though he did not intend it, and may not even have expected it, he brought another obscure theory of the origin of life to prominence as well.

In the 1830s, Berzelius studied a piece of meteorite that was found near Alais, France in 1806. Berzelius analyzed the Alais meteorite piece and found the chemical composition to be some unknown sublimate, ammonia, and other elements. He was surprised to find a blackish substance among those elements. That blackish substance had the complex carbon compounds, which had significant meaning for Berzelius. Being a vitalist, he believed that the organic compounds couldn't be generated from the inorganic matter. Following that theory, finding the carbon compounds in a meteorite would point at the possibility of the presence of organic matter or life outside the Earth. But Berzelius did not believe that was the case. He mentioned that the properties of the carbon substance in the meteorite were very different to what we find on Earth. Hence, we cannot conclude that organic matter exists wherever the meteor came from.

On 13 October 1838, a meteorite fell in Bokkeveld, near Cape Town. And on 15 April 1857, another meteorite fell in Kaba, Hungary. Berzelius's student Wohler, who is known for Wohler's synthesis, undertook a chemical analysis of these meteorite rocks and published the results on the Kaba (1857) and Cape (1838) meteorites in the 1850s.

Wohler found that both the meteorites were remarkably similar in their composition. He confirmed that both had the carbon substances and organic origins. In his paper "On the composition of the Cape meteorite," Wohler says,

> *On account of the small quantity of the material, it was impossible to examine it more accurately; but there is no doubt that this meteoric mass, falling, as it did, to the earth from planetary space, contains a carbonaceous substance which can have no other than an organic origin That the almost black colour of the due intimately.*

Panspermia

The theory that there is life throughout the universe is called "panspermia." This concept of panspermia is important to the subject of the origin of life because it hints at the possibility of life, or organic matter, reaching the Earth from space.

The "plurality of worlds" also suggests the similar possibility. It states there are many worlds in the universe just like Earth that harbor life. The history of this idea goes back to Anaximander. Later, Giordano Bruno (1548–1600) was a champion of the "plurality of worlds" idea, and his conviction about this idea led to him being burned at stake for heresy.

Neither panspermia nor "plurality of worlds" fully explains how the life originated on Earth. Also, to clarify, even though Pasteur, Tyndall and others were sometimes called the "panspermists," it was referred to their support for the idea that the life (or germs) was present in the atmosphere.

Richter

The first person to come up with the hypothesis of panspermia in its modern sense was a reputed German physician called Hermann Eberhard Friedrich Richter (1808- 1876) from Leipzig. Influenced by an 1862 book *La pluralité des mondes habités* (The Plurality of Inhabited Worlds) written by a French astronomer and author Camille Flammarion (1842-1925), and reading Berzelius's findings of the Alais meteorite, Richter came up with the idea of life being distributed across the universe via meteorites and comets. He noticed that a meteorite entering the Earth's atmosphere is not free from the contamination. When it enters the Earth's atmosphere, it picks up the minute particles in the air, including living germs, and carries them with it to the ground. Similarly, he suggests that a meteorite could pick up living matter from the space and carry it along as it enters the Earth. This hypothesis, later termed as "*lithopanspermia,*" says that the moving rocks like the meteorite carry the living matter from one space object to another.

Helmholtz

A few years later, another eminent German scientist supported the panspermia hypothesis—Hermann Ludwig Ferdinand von Helmholtz (1821–1894). This is the same Helmholtz who came up with a filtration experiment in 1843 showing that the fermentation is not a "purely" chemical process, as Liebig had suggested.

Helmholtz was a physicist as well as a physician. He could apply the laws of physics in the studies of physiology. Though a student of Muller at the Friedrich Wilhelm Medical Institute in Berlin, he did not support the theory of vitalism as his tutor did. After graduation, he had to serve as an army doctor in return for the free medical education. Then he joined the Prussian University of Königsberg as an associate professor of physiology in 1849. But even before the start of his career, Helmholtz had come up with the theories of conservation of force and the conservation of energy while working on the animal heat, thus opposing the vital force theory. He invented the ophthalmoscope in 1851 to assist in eye examination. And at a later stage of his career, he concentrated completely on electrodynamics.

Helmholtz expressed his support for the lithopanspermia theory in a public lecture in 1871 –

> *Who can say whether the comets and meteors ... may not scatter germs of life wherever a new world has reached the stage in which it is a suitable dwelling place for organic beings? We might, perhaps, consider such life to be allied to ours, at least in germ, however different the form it might assume in adapting to its new dwelling place.*

A few months later a close friend of Helmholtz, William Thomson (known later as Lord Kelvin), also came forth supporting the hypothesis. The sudden support for the panspermia theory from such eminent scientists of the time demanded attention to the subject.

Lord Kelvin

William Thomson (1824-1907) studied under the supervision of his father James Thomson, a teacher of Mathematics at the Royal Belfast Academical Institution and later at the University of Glasgow. William Thomson was a brilliant student, winning multiple prizes and publishing the scientific papers by the time he was 18. He would follow the ongoing research projects and scientific investigations and add contributions from his end. He completed his graduation from Cambridge by the time he was 21, and in another year he was appointed as a professor of natural philosophy at the University of Glasgow. He would hold that position until the end of his professional career.

Thomson had made a name for himself in science at a very

young age. Often he collaborated with the other scientists on the works they had already started, for example, he worked on some theories of Faraday and Joule. While doing so, he contributed to the first and second laws of thermodynamics. Thomson became famous in public after his venture as an engineer delivering the first successful transatlantic telegraph cable. He also came up with the absolute temperature scale that would be known as Kelvin scale in his honor. Later "kelvin" (K) became one of the base units in the 'International System of Unit' system and used to measure absolute temperature.

The Cockshy Theory

A couple of years before his venture as an engineer with the Atlantic Telegraph Company, Thomson read Helmholtz's paper on the conservation of energy and was elated. He wanted to meet Helmholtz in person, so he sent him an invite to attend the BAAS meeting. Helmholtz accepted, and thus began a friendship between these two scientists that would only continue to grow as time passed. So it is not much of a surprise that both came out in support of the lithopanspermia theory around the same time.

Huxley had given the presidential address at BAAS in 1870, in which he clarified his stand on spontaneous generation. At the next BAAS annual meeting held on 2nd August 1871, Thomson had the opportunity to give the presidential address, and he discussed the origin of life and evolution in his speech. Thomson rejected Darwin's theory of evolution by natural selection. He strongly favored the idea of some "intelligence" at play in the world around us. In the speech, he refers to Sir John Herschel's criticism of the natural selection for "*it did not sufficiently take into account a continually guiding and controlling intelligence.*" Supporting that criticism Thomson says,

> *...overpoweringly strong proofs of intelligent and benevolent design lie all round us, and if ever perplexities, whether metaphysical or scientific, turn us away from them for a time, they come back upon us with irresistible force, showing to us through nature the influence of a free will, and teaching us that all living beings depend on one ever-acting Creator and Ruler.*

There was another reason why Thomson rejected natural selection. He had firmly held that the age of the Earth is definite and not old enough to have allowed enough time for evolution to happen through the process that Darwin had suggested. Until that time, the problem of the age of Earth belonged to geology, but Thomson thought it was not possible to measure the rates of changes in geological processes. So instead, he pushed the problem to physics and came up with a formula based on the temperature gradient. The parameters of the formula still depended on some uncertain parameters, like the temperature at the core of the Earth. So he estimated a range of values for these parameters based on different calculations and then used the formula to calculate the age range of the Earth. Going by these, Thomson's estimation of the age of the Earth was somewhere between 20 to 400 million years, which he would later update to 20 to 40 million years. But Thomson argued that even the upper limit of 400 million years was not enough for the evolution to have occurred in the way Darwin suggested, because it included the years when the Earth was just a hot globe where no life could survive.

Thomson rejected the natural selection theory of evolution, but he was not against the concept of evolution altogether. He rejected none of the evolution theories on religious grounds alone. He was rather considerate to the idea of evolution and agreed in the speech that all the life forms we observe today are most likely the result of an evolution process:

> *From the Earth stocked with such vegetation as it could receive meteorically, to the Earth teeming with all the endless variety of plants and animals which now inhabit it, the step is prodigious; yet, according to the doctrine of continuity, most ably laid before the Association by a predecessor in this Chair (Mr. Grove), all creatures now living on earth have proceeded by orderly evolution from some such origin.*

Thomson also had a clash with Huxley in 1869 as Huxley had challenged his calculations related to the cooling of the early Earth. But he supported Huxley's opinion regarding the spontaneous generation in his speech. He referred to Huxley's presidential speech and dismissed spontaneous generation:

> *...science brings a vast mass of inductive evidence against this hypothesis of spontaneous generation, as you have heard from my predecessor in the Presidential chair. Careful enough scrutiny has, in every case up to the present day, discovered life as antecedent to life. Dead matter cannot become living without coming under the influence of matter previ-*

ously alive. This seems to me as sure a teaching of science as the law of gravitation. ... I confess to being deeply impressed by the evidence put before us by Professor Huxley, and I am ready to adopt, as an article of scientific faith, true through all space and through all time, that life proceeds from life, and from nothing but life.

If spontaneous generation is not the correct answer, then how did life on Earth begin? Thomson reiterated Helmholtz in answer to the question. But why didn't Thomson consider the creator's role in the origin of life? As he tells, "*If a probable solution, consistent with the ordinary course of nature, can be found, we must not invoke an abnormal act of Creative Power.*" And that is what he tried to do with the suggestion of panspermia.

Thomson placed the example of lava erupting from the volcano, which erupts, cools down, and then life begins on that land. And the reason for life to begin on such a land is that the seeds of life traveling through the air or water must have reached it. Thomson extended the same explanation in the case of early Earth. Once the Earth cooled down and was ready to support life, some meteorite brought seeds of life with it. Those seeds splattered when the meteorite entered Earth's atmosphere thus paving the way for the beginning of life on Earth. But Thomson's proposal of meteors containing the seeds of life came from the fact that he had taken the "plurality of worlds" for granted:

...we all confidently believe that there are at present, and have been from time immemorial, many worlds of life

besides our own, we must regard it as probable in the highest degree that there are countless seed-bearing meteoric stones moving about through space.

The possibility of the origin of life from lithopanspermia was not unscientific. As Thomson put it:

> *The hypothesis that life originated on this Earth through moss-grown fragments from the ruins of another world may seem wild and visionary; all I maintain is that it is not unscientific.*

In reaction, most scientists were astounded at his claims, and many criticized him. Hooker wrote to Darwin on 5 August 1871 as soon as he read Thomson's presidential address, expressing displeasure over Thomson's remarks about natural selection and his life-from-meteors hypothesis:

> *The notion of introducing life by meteors is astounding & very unphilosophical, as being dragged in head & shoulders apropos of the speculations of the "Origin" of life from or amongst existing matter—seeing that Meteorites are after all composed of the same matter as the globe is—5 Does he suppose that God's breathing upon Meteors or their progenitors is more philosophical than breathing on the face of the earth?*

Darwin replied to Hooker the very next day concurring with him: "*...have read with greatest interest Thompson's address: but you say so exactly & fully all that I think, that you* *have taken all the words from my mouth.*" In his next letter to Darwin on 15 August 1871, Hooker also indicated that even Huxley had criticized Thomson's hypothesis in the biology section of the BAAS meeting–:

> *Huxley rates Thompson's the "Cock-shy" theory— God makes a cockshy of the world. I hear that he baisted T. awfully in section D.*

Thomson knew the criticism he would receive for supporting panspermia, but it did not stop him. He continued to push for studies of meteorite rocks in search of an indication of life. He firmly believed that all the objections to panspermia could be overcome. One objection was that the germs in the meteorite would burn due to the extreme heat generated when the meteorite enters Earth's atmosphere. But both Thomson and Helmholtz answered that the seeds of life in the meteorite would spatter in the atmosphere as soon as it began its entry. Another objection was based on religion. Many thought that the lithopanspermia theory suggested that God had no role in creation. Thomson clarified his position on this by maintaining that he does not deny the role of any creative power. Instead, he was suggesting that some intelligence must have provided us with a natural process for the origin of life, which could be explained by science.

The theory of panspermia mostly belonged to philosophy until Richter and Helmholtz brought it to the scientific community. Thomson was far more successful than his predecessors in bringing the theory to the attention of the scientific world as a possible alternate theory for the origin of life on

Earth. It was good enough for like-minded future scientists to consider it seriously and put efforts towards finding the evidence. Panspermia was still not a scientific theory as there was no evidence to support it, but it had moved from being a random idea to a scientific hypothesis that followed a logical scientific explanation.

If we extend this hypothesis and try to fill the gap between the origin of life on Earth and the origin of man, we understand that panspermia depends on evolution. The seeds of life reaching the Earth from space through any medium such as meteorite can be only a very basic form of life. Complex life forms, such as humans, could not have possibly come from the space directly. The seeds of life, the nature of which was still unclear, might either be the smallest organisms or some form of organic matter. So, how did these result into a vast different multitude of life forms we see all around us? At this point, evolution is the only answer available to it. In his presidential speech, Thomson accepted this as a probable case.

False Positive

Many false positives follow any new groundbreaking theory, and this happened even in the case of the lithopanspermia theory. In the 19th century, many studies were done on the different meteorite rocks trying to find evidence of some organic life but with no success. In 1879, Dr. Otto Hahn (1828–1904), a former German lawyer and an amateur geologist, thought he found the evidence for the origin of life on Earth.

Otto Hahn studied the Knyahinya meteorite that fell in

Hungary in 1866. He was ecstatic when he noticed what he thought it to be organic fossils in the meteorite. He published the results in his 1879 book *Die Urzelle.* He even sent the book to Darwin as a token of respect. In 1880, he wrote another book called *Die Meteorite und ihre Organismen* in which he published better photographs of the fossils of sponges, corals, and crinoids he had found in the meteorite. He also sent the findings to a German zoologist David F. Weinland, who supported Hahn's claims and even added his own observations.

It did not take long for others to come forward and dismiss Hahn's findings as products of his imagination. They confirmed and then criticized him for failing to differentiate between organic structures and minerals. But much later, on 27 August 1881, the journal *Science* published an interesting article that said Darwin had corresponded with Otto Hahn praising him for important findings. The article claimed that Darwin had replied to Otto Hahn upon reading his work saying,

> *...it seems to be very difficult to doubt that your photographs exhibit organic structure...*
>
> *...your discovery is certainly one of the most important.*

Further, the article suggested that Otto Hahn visited Darwin to show his specimens, and after checking one fine specimen under the microscope, Darwin apparently "*started up from his seat and exclaimed*":

Almighty God! What a wonderful discovery! Wonderful!

...Now reached life down!

But not believing this story, Thomas George Bonney (1833–1923), a British professor of geology, followed up with Francis Darwin and confirmed that the story was not true. In any case, it did not take long to discredit Otto Hahn's false positive find.

So by the end of 19th century, there was no evidence for the theory of panspermia. If there were any evidence for panspermia, then it would force us to look at a different direction to find an answer to the origin of life. It would answer how life began on Earth, but it would pose an even bigger question – how did life begin in the universe?

EPILOGUE

Our efforts in the quest for the origin of life on Earth began with an inquisitive mind. The ancient, questioning mind, with no concept of science, tried to use the logic based on naked-eye observations. It would be hard to determine when exactly the philosophy and science fields were born, since the meaning of these terms has changed greatly over the time. But perhaps we can say with some confidence that when a man tried to come up with the rational explanations based on some evidence that others could reproduce, at that point science was born. In Greek history, this kind of work was done by Hippocrates in medicine and Aristotle in science. They laid solid ground for science, philosophy, and medicine that coming generations rarely dared to question. And when there are no new questions there will be no new answers.

Aristotle and Hippocrates were finally questioned after the Renaissance in Europe, thus leading to new discoveries, inventions, and changes in the fields of science, medicine, and philosophy. Innovations resulted in better tools, and the tools helped us achieve great results. One such example was the microscope, which opened the gates to a new world of invisible organisms that live all around us. The course of the history of science changed substantially after that.

The 18th and 19th centuries have perhaps seen as many breakthroughs in the scientific and philosophy fields as all the previous centuries did together. In the field of philosophy, the struggle between the idealists and materialists intensified. Based on their ideologies, philosophers differed in their support for the origin of life on Earth by natural means. In science, we see developments in many related concepts such as vitalism, fermentation theories, atmospheric germ theory, and disease theory.

The 19th century culminated with Pasteur destroying the credibility of spontaneous generation, and Tyndall successfully putting the matter to rest once and for all. Panspermia and lithopanspermia theories had just entered the scientific circle but with absolutely no evidence to support them. Abiogenesis was still an open question, but it had seen a setback due to its relationship to heterogenesis. The investigators of the problem were in desperate need for new direction to look for a solution. At this point, the future of the pursuit for the origin of life on Earth deemed hopeless for the majority.

What Next?

As the 19th century gave way to the 20th, natural selection slowly gained wider acceptance. It succeeded in converting the majority of scholars in a relatively short time. Evolution made the problem of the origin of life on Earth simpler. There was no need to investigate how all life forms came into existence. All one had to do was to figure out how the very first life on Earth came into being. Spontaneous generation was no more an answer to this, but abiogenesis was. Abiogenesis suggested the generation of life from non-life, and it must have happened only in the case of the very first life form. In the early Earth, once the environment was suitable for the first life, different essential chemical ingredients must have come together through a natural process to form a simple single-celled life form. To reiterate Charles Darwin's hopes in a letter to Hooker on 1 February 1871:

> *It is often said that all the conditions for the first production of a living organism are now present, which could ever have been present.— But if (& oh what a big if) we could conceive in some warm little pond with all sorts of ammonia & phosphoric salts,—light, heat, electricity &c present, that a protein compound was chemically formed, ready to undergo still more complex changes, at the present day such matter would be instantly devoured, or absorbed, which would not have been the case before living creatures were formed.*

This gave hope and direction for the future researchers to investigate the exact mechanisms that would have led to the origin of life on Earth.

A better understanding of the cell and the cell components, such as the structure and function of DNA and RNA, pushed the research to new extremes. To understand how the first life must have formed, it is vital to understand how the first building blocks of life could have formed. Once these building blocks formed, the next big question is to figure how they replicated and reproduced. There are efforts seen since the beginning of 20th century.

In parallel, panspermia gained acceptance from many modern philosophers and researchers. As mankind turned its eyes toward the heavens and finally ventured beyond the confines of our atmosphere, the possibility of finding evidence of our extraterrestrial origins seemed more real than ever. There were few exciting moments for the people when they heard the news of life matter found in space, only to find later that they were false positive cases. While in some cases researchers continue to argue that their findings conclusively prove the existence of life in space, the majority of the science community is not willing to accept them. Whether there is a conclusive evidence available or not, one cannot deny that panspermia theory gained popularity in the 20th century.

This gives a brief idea of what we expect in the next part. Before concluding, I would like to put forward few questions for the reader to consider:

- We can understand why man conceived of the idea of spontaneous generation. But, even after the scientific method was introduced and the microorganisms were discovered, why did it take so long to consider the new evidence in front of us? We can apply the same question to other theories such as vitalism, humorism, and Miasmata.
- Pasteur's religious beliefs stood against spontaneous generation, but Pouchet's religious beliefs found the creator's act in it. Why do we see such contradiction?

- Semmelweis came up with excellent experimental results, yet all the scholars rejected it without even considering a trial of his work. Was it only because of the lack of a theoretical support to the experimental outcome?
- Pasteur emerged on the winning side of history. But,why was he convinced that the negative results he obtained in the swan neck flask experiments were faulty?
- Why was Darwin criticized heavily for not discussing the origin of life problem in his theory of evolution? Does leaving the origin of life as an open question makes any difference to the credibility of his theory?
- Did Bastian lose focus by ignoring the developments in the germ theory and its applications? If he did ignore the developments around him, why was that?

These questions lead to a discussion, or debate, in our minds. They deal with the human nature, differences in philosophical outlooks, preconceived notions, and individual want for a complete story, with or without scientific evidence.

BIBLIOGRAPHY

Aristotle. 2017. *On the Generation of Animals.* Translated by Arthur Platt. The University of Adelaide, 03 31.

Aristotle. 2017. *The History of Animals.* Translated by D'Arcy Wentworth Thompson. The University of Adelaide, 03 31.

Bastian, H. Charlton. 1870. "Facts and reasonings concerning the heterogenous evolution of living things; Part I of III." *Nature* (Macmillan Journals) 170-177.

Bastian, H. Charlton. 1870. "Facts and reasonings concerning the heterogenous evolution of living things; Part II of III." *Nature* (Macmillan Journals) 193-201.

Bastian, H. Charlton. 1870. "Facts and reasonings concerning the heterogenous evolution of living things; Part III of III." *Nature* (Macmillan Journals) 219-228.

Bastian, H. Charlton. 1870. "Letters to the editor." *Nature* (Macmillan Journals) 492.

Bastian, H. Charlton. 1870. "Reply to Professor Huxley's inaugural address at Liverpool on the question of the origin of life; Part I of II." *Nature* (Macmillan Journals) 410-413.

Bastian, H. Charlton. 1870. "Reply to Professor Huxley's inaugural address at Liverpool on the question of the origin of life; Part II of II." *Nature* (Macmillan Journals) 431-434.

Bastian, Henry Charlton. 1872. *The beginnings of life, being some account of the nature, modes of origin and transformations of lower organisms.* London: Macmillan and Co.

Browne, Thomas, Sir,. 1658. *Pseudodoxia epidemica : or, enquiries into very many received tenents, and commonly presumed truths.* London: printed for Edward Dod, and are to be sould by Andrew Crook.

Buchner, Ludwig. 1884. *Force and Matter or Principles of the Natural Order of the Universe. With a System of Morality based Thereon.* 4th English Ed. London: Asher and Co.

Clendening, Logan. 1937. "The Doctor Abroad." *Hygeia* (American Medical Association) 15 (1-6): 338-341, 347.

Crowe, Michael J. 1988. *The Extraterrestrial Life Debate 1750-1900: The Idea of a Plurality of Worlds from Kant to Lowell.* illustrated, reprint. Cambridge: Cambridge University Press.

Cuvier, Georges. 1836. "Biographical Memoir of M. de Lamarck." *Edinburgh New Philosophical Journal, Volume 20*(Adam and Charles Black) 1-22.

Darwin, Charles. 1859. *On the origin of species by means of natural selection, or, The preservation of favoured races in the struggle for life.* 1st Edition. London: John Murray.

Darwin, Charles. 2008. *The Autobiography of Charles Darwin From The Life and Letters of Charles Darwin.* Edited by Francis Darwin. Prods. Sue Asscher and David Widger. 09 21.

—. 1887. *The Life and Letters of Charles Darwin: Including an Autobiographical Chapter Vol 1-3.* Edited by Francis Darwin. London: John Murray.

Dubos, René. 1986. *Louis Pasteur, Free Lance of Science.* illustrated, reprint, revised. New York: Da Capo Press.

Engelhardt, Hugo Tristram, and Arthur L. Caplan, . 1987. *Scientific Controversies: Case Studies in the Resolution and Closure of Disputes in Science and Technology.* illustrated, reprint. New York, NY: Cambridge University Press.

Engels, Frederick. 1934. *Dialectics of Nature.* Translated by Clemens Dutt. Progress Publishers.

Gregory, Mary Efrosini. 2008. *Evolutionism in Eighteenth-century French Thought.* New York: Peter Lang.

Haeckel, Ernst. 1880. *The History of Creation, Vol. I (of 2) Or the Development of the Earth and its Inhabitants by the Action of Natural Causes.* Translated by Edwin Ray Lankaster. Vol. I. II vols. New York: D. APPLETON AND COMPANY.

—. 1880. *The History of Creation, Vol. II (of 2) Or the Development of the Earth and its Inhabitants by the Action of Natural Causes.* Translated by Edwin Ray Lankaster. Vol. II. II vols. New York: D. APPLETON AND COMPANY.

Helmont, John Baptista Van. 1664. *Van Helmont's Workes, containing his ... Philosophy, Physick, Chirurgery, Anatomy; wherein the philosophy of the schools is examined, the errours therein refuted ..., done into English by J. C. [i.e. J. Chandler,] sometime of M. H. Oxon.* London: The British Library. https://books.google.com/books?id=JxZmAAAAcAAJ.

Hooke, Robert. 2005. *Micrographia: Some Physiological Descriptions of Minute Bodies Made by Magnifying Glasses with Observations and Inquiries Thereupon.* Prods. Robert Shimmin and Keith Edkins. March 29.

Hutton, Charles, George Shaw, and Richard Pearson, . 1809. *The Philosophical Transactions of the Royal Society of London from their commencement, in 1665, to the year 1800; Abridged, with notes and biographic illustrations.* Vol. I. London: C. AND R. BALDWIN.

Huxley, Thomas Henry. 1870. "Letters to the editor - Dr. Bastian and Spontaneous Generation." *Nature* (Macmillan Journals) 473.

—. 1913. *Life and Letters of Thomas Henry Huxley.* 2. Edited by Leonard Huxley. Vol. I. III vols. London: Macmillan and Co.

—. 1913. *Life and Letters of Thomas Henry Huxley.* Edited by Leonard Huxley. Vol. III. III vols. London: Macmillan and Co.

—. 1903. *Life and Letters of Thomas Henry Huxley.* 2. Edited by Leonard Huxley. Vol. II. III vols. London: Macmillan and Co.

Huxley, Thomas Henry. 1870. "The British Association. Liverpool meeting, 1870." *Nature* (Macmillan Journals) 399-408.

Koenigsberger, Leo. 1906. *Hermann von Helmholtz.* Translated by Frances A. Welby. Oxford: THE CLARENDON PRESS.

Maillet, Benoît de. 1750. *Telliamed, or, The world explain'd : containing discourses between an Indian philospher and a missionary, on the diminution of the sea, the formation of the earth, the origin of men & animals : and other singular subjects, relating to natural history & philosphy ; a very curious work.* London: Printed for T. Osborne.

Marx, Karl. 1932. *Economic and Philosophic Manuscripts of 1844.* Translated by Martin Mulligan. Moscow : Progress Publishers.

McCartney, Eugene S. 2017. "Spontaneous Generation and Kindred Notions in Antiquity." *Transactions and Proceedings of the American Philological Association, Vol 51 (1920).* The Johns Hopkins University Press, 04 02.

Meyer, Arthur William. 1936. *An Analysis of the De Generatione Animalium of William Harvey.* Stanford: Stanford University Press.

Mitchell, Peter Chalmers. 1900. *Thomas Henry Huxley: a sketch of his life and work.* New York: G.P. Putnam's sons.

1881 "MR. DARWIN ON DR. HAHN'S DISCOVERY OF FOSSIL ORGANISMS IN METEORITES." *Science* (John Michels) 2 (62).

Oken, Lorenz. 1847. *Elements of Physiophilosophy.* Translated by Alfred Tulk. London: Ray Society.

Redi, Francesco. 1909. *Experiments on the generation of insects.* Edited by Mab Bigelow and Robert Payne Bigelow. Translated by Mab Bigelow. The Open court publishing company.

Rennie, James. 1838. *Insect transformations.* London: C. Knight and Co.

Strick, James, ed. 2001. *Evolution and the Spontaneous Generation Debate.* Vol. I. Bristol: Thoemmes Press.

Thomson, William Sir. 1871. "The British Association meeting at Edinburgh." *Nature* 261-278.

2016 02 15 Accessed 2016 02 15 http://www.pasteurbrewing.com/

1809 *TRANSACTIONS OF THE AMERICAN PHILOSOPHICAL SOCIETY, HELD AT PHILADELPHIA, FOR PROMOTING USEFUL KNOWLEDGE.* Vol. VI. PHILADELPHIA: C. AND A. CONRAD AND CO.

Tyndall, John. 1888. *Essays on the floating-matter of the air, in relation to putrefaction and infection.* New York: D. Appleton and Company.

—. 1900. *Fragments of science; Part I and II.* New York: P.F. Collier.

Voltaire. 1901. *Voltaire's Philosophical Dictionary: Unabridged and Unexpurgated, Volume 5.* E. R. DuMont.

Wöhler, Friedrich. 1859. "On the Composition of the Cape Meteorite." In *The London, Edinburgh, and Dublin Philosophical Magazine and Journal of Science; Volume XVIII Series IV*, edited by David Brewster, Robert Kane, William Francis and John Tyndall, 213-218. London: Taylor and Francis.

Webster, Charles, ed. 2012. *The Intellectual Revolution of the Seventeenth Century (Routledge Revivals).* New York: Routledge.

Wyman, Jeffries. 1867. *Observations and Experiements on Living Organisms in Heated Water.* NEW HAVEN: M. M. BURDICK.

Made in the USA
Monee, IL
04 November 2026

67387471R10144